# The IMA Volumes in Mathematics and its Applications

## Volume 98

*Series Editors*
Avner Friedman Robert Gulliver

**Springer**
*New York*
*Berlin*
*Heidelberg*
*Barcelona*
*Budapest*
*Hong Kong*
*London*
*Milan*
*Paris*
*Santa Clara*
*Singapore*
*Tokyo*

# Institute for Mathematics and its Applications
# IMA

The **Institute for Mathematics and its Applications** was established by a grant from the National Science Foundation to the University of Minnesota in 1982. The IMA seeks to encourage the development and study of fresh mathematical concepts and questions of concern to the other sciences by bringing together mathematicians and scientists from diverse fields in an atmosphere that will stimulate discussion and collaboration.

The IMA Volumes are intended to involve the broader scientific community in this process.

Avner Friedman, Director

Robert Gulliver, Associate Director

* * * * * * * * * * *

## IMA ANNUAL PROGRAMS

| | |
|---|---|
| 1982–1983 | Statistical and Continuum Approaches to Phase Transition |
| 1983–1984 | Mathematical Models for the Economics of Decentralized Resource Allocation |
| 1984–1985 | Continuum Physics and Partial Differential Equations |
| 1985–1986 | Stochastic Differential Equations and Their Applications |
| 1986–1987 | Scientific Computation |
| 1987–1988 | Applied Combinatorics |
| 1988–1989 | Nonlinear Waves |
| 1989–1990 | Dynamical Systems and Their Applications |
| 1990–1991 | Phase Transitions and Free Boundaries |
| 1991–1992 | Applied Linear Algebra |
| 1992–1993 | Control Theory and its Applications |
| 1993–1994 | Emerging Applications of Probability |
| 1994–1995 | Waves and Scattering |
| 1995–1996 | Mathematical Methods in Material Science |
| 1996–1997 | Mathematics of High Performance Computing |
| 1997–1998 | Emerging Applications of Dynamical Systems |
| 1998–1999 | Mathematics in Biology |
| 1999–2000 | Reactive Flows and Transport Phenomena |

Continued at the back

Donald A. Drew    Daniel D. Joseph
Stephen L. Passman
Editors

# Particulate Flows

## Processing and Rheology

With 12 Illustrations

Springer

Donald A. Drew
Department of Mathematical Science
Rensselaer Polytechnic Institute
Troy, NY 12180, USA

Daniel D. Joseph
Department of Aerospace Engineering and Mechanics
University of Minnesota
Minneapolis, MN 55455, USA

Stephen L. Passman
Sandia National Laboratory
Albuquerque, NM 87185, USA

*Series Editors:*
Avner Friedman
Robert Gulliver
Institute for Mathematics and its Applications
University of Minnesota
Minneapolis, MN 55455, USA

---

Mathematics Subject Classifications (1991): 76A05, 76A10, 76B15, 76D05, 82B21, 82B44, 82C22, 82C44, 82C70, 82D30

---

Library of Congress Cataloging-in-Publication Data
Particulate flows : processing and rheology / [edited by] Donald A. Drew, Daniel D. Joseph, Stephen L. Passman.
p. cm. – (The IMA volumes in mathematics and its applications ; 98)
Includes bibliographical references.
ISBN-13: 978-1-4684-7111-3 e-ISBN-13: 978-1-4684-7109-0
DOI: 10.1007/978-1-4684-7109-0
1. Inhomogeneous materials – Fluid dynamics – Congresses. 2. Bulk solids flow – Mathematical models – Congresses. 3. Granular materials – Fluid dynamics – Congresses. I. Drew, Donald A. (Donald Allen), 1945– . II. Joseph, Daniel D. III. Passman, Stephen L. IV. Series. V. Series: IMA volumes in mathematics and its applications ; v. 98.
TA418.9.I53P37 1997
531′.1134–dc21 97-41795

Printed on acid-free paper.


Softcover reprint of the hardcover 1st edition 1998


Production managed by Karina Mikhli; manufacturing supervised by Jeff Taub.
Camera-ready copy prepared by the IMA.

9 8 7 6 5 4 3 2 1
ISBN-13: 978-1-4684-7111-3

## FOREWORD

This IMA Volume in Mathematics and its Applications

## PARTICULATE FLOWS: PROCESSING AND RHEOLOGY

is based on the proceedings of a very successful one-week workshop with the same title, which was an integral part of the 1995–1996 IMA program on "Mathematical Methods in Materials Science." We would like to thank Donald A. Drew, Daniel D. Joseph, and Stephen L. Passman for their excellent work as organizers of the meeting.

We also take this opportunity to thank the National Science Foundation (NSF), the Army Research Office (ARO) and the Office of Naval Research (ONR), whose financial support made the workshop possible.

Avner Friedman

Robert Gulliver

# PREFACE

The workshop on Particulate Flows: Processing and Rheology was held January 8–12, 1996 at the Institute for Mathematics and its Applications on the University of Minnesota Twin Cities campus as part of the 1995–96 Program on Mathematical Methods in Materials Science. There were about forty participants, and some lively discussions, in spite of the fact that bad weather on the east coast kept some participants from attending, and caused scheduling changes throughout the workshop.

Heterogeneous materials can behave strangely, even in simple flow situations. For example, a mixture of solid particles in a liquid can exhibit behavior that seems solid-like or fluid-like, and attempting to measure the "viscosity" of such a mixture leads to contradictions and "unrepeatable" experiments. Even so, such materials are commonly used in manufacturing and processing.

Food processing, catalytic processing, slurries, coating, paper manufacturing, particle injection molding, paving and filter operation all involve flow of heterogeneous materials. In many of these processes, the rheology of such materials is a critical element in considerations of design, operation, and efficiency. Also, in many of them, the properties are nonuniform in space and change in time. Consequently, using these materials represents a technological challenge.

A dilemma facing a researcher who must analyze a dispersed particulate flow is that a description of the motions of the individual particles and the surrounding fluid is almost within reach of modern computers and numerical solution routines. Numerical simulation of the motion of a few hundred rigid particles now appear in the literature, but the situation with deformable dispersed units (bubbles or droplets) is still limited to a few such units. Consequently, models that treat the dispersed and continuous phase as continua have been developed and used since the 1950's. These models are adequate in some situations. It appears that those which have close associations between the micro dynamics (that is, dynamics on the particle scale) and the interactions and stresses that are responsible for the rheology seem to do better.

One of the fundamental rheological issues is the microstructure and its evolution. Numerical solutions and experiments both showed the importance of structures in the flow. A judicious mix of physical understanding and mathematical analysis is needed to understand the rheology of particle-fluid mixtures. The solutions of the equations of motion show some of the richness of the phenomena encountered.

The papers included in this volume connect physical phenomena with analysis of microscale dynamics and particle distribution functions to build a better understanding of the physics and to develop mathematical models of the phenomena, and provide analyses of the resulting partial differential equations.

We would like to express our thanks and those of all the participants to Avner Friedman and Robert Gulliver of the IMA for their help in organizing and hosting the Workshop. It is also a pleasure to thank the IMA staff, especially Amy Cavanaugh, for their very special assistance in the running of the Workhop and Patricia V. Brick for her usual expert assistance in the production of these proceedings.

Donald A. Drew

Daniel D. Joseph

Stephen L. Passman

June 1997

## CONTENTS

# COMPRESSIBLE FLOW OF GRANULAR MATERIALS

TOMMASO ASTARITA*, RAFFAELLA OCONE†, AND GIANNI ASTARITA‡

**Abstract.** The analysis of compressible flow of granular materials is considered. The necessary thermodynamic background has recently been developed, and this is reviewed from the viewpoint of compressible flow theory. Elementary problems can be solved rather easily following traditional methods of classical gas dynamics. However, even moderately complex problems present subtle unexpected difficulties, which are shown to be related to the inelasticity of particle-particle collisions.

**1. Introduction.** Granular systems are constituted by a large number of solid particles dispersed in an interstitial fluid. Suspensions and colloidal systems of course are in the same category, but the words "granular systems" are commonly used to refer to systems where a), surface forces (such as electrostatic and Van der Waals forces) do not play a major role, and b), the relative average velocity between particles and interstitial fluid is significant.

One of the approaches to the analysis of the flow of granular systems is generally based on what may be called two-phase models. The interstitial fluid and the assembly of particles are regarded as two equivalent continuum phases which interpenetrate (both phases occupy simultaneously the same volume). Two-phase models are of necessity based on regarding local average values as point values in the equivalent continuum: to see this, one only needs to realize that even the concept of a local solid volume fraction $\epsilon$ is defined only in an average sense. Let $d_p$ be the (average) particle diameter. Local averages, indicated with triangular brackets in the following, are intended as averages over a volume of linear dimension $\ell$, with, of course, $\ell >> d_p$: the value of, for instance, $\epsilon$ is well defined over such a volume. Since $\epsilon$ is regarded as a point value when the appropriate equations are written down for the equivalent continuum, the scale of description of the whole system, $L$, needs to be well in excess of $\ell$, $L >> \ell$.

Starting from the pioneering work of Anderson and Jackson (1967), the question of how to correctly define averages is well discussed in the literature (Bowen 1971; Drew 1983; Joseph et al. 1990; Lundgren 1972), and we do not discuss this here, except to state that the requirement for any appropriate averaging procedure is that the exact solution of the averaged equations should coincide (in some sense to be made precise) with the average solution of the exact equations. Let $\phi_p$ be the (constant) particle density. The local average density of the particulate phase, $\phi$, is of course

* Department of Energetics, Thermofluodynamics and Environmental Control, Universita' di Napoli "Federico II", Piazzale Tecchio, 80125 Napoli, Italy.

† Department of Chemical Engineering, University of Nottingham, University Park, Nottingham NG7 2RD, UK.

‡ Department of Materials and Production Engineering, Universita' di Napoli "Federico II", Piazzale Tecchio, 80125 Napoli, Italy.

$\epsilon\phi_p$. It follows that, even though $\phi_p$ is constant, the particulate phase is "compressible" in that $\epsilon$ may very well not be constant in both space and time. In this paper attention is focussed on the analysis of the flow of granular systems under conditions where the compressibility of the particulate phase is of relevance.

This paper is largely expository in nature, in that recent results on the granular thermodynamic theory and its application to elementary compressible flow problems are critically reviewed in sections 2-4. However, section 5 presents some original results on moderately complex compressible flow problems; these results are only preliminary in character, in that conceptual difficulties are identified but no solution of them is presented.

**2. Large scale thermodynamics.** Compressible flow is a well developed subject in gas dynamics. Classical compressible flow theory is rooted on a firm background of thermodynamics, and an analogous background is required if one wants to develop the theory of compressible flow of granular materials. In classical statistical thermodynamics, averages are taken over a "local" scale which is required to be large with respect to molecular mean free path dimensions. What is needed for granular materials is a large scale statistical theory of thermodynamics based on averages at the $\ell$-scale, where solid particles play the role that molecules play in the Maxwellian theory. In particular, large scale or granular analogs of classical thermodynamic quantities such as temperature, entropy, enthalpy, etc. need to be available.

The concept of granular temperature, which was first explicitly introduced by Ogawa (1978), but can in fact be traced back to the work of Bagnold (1954), has been widely used in the literature on granular systems. Let $\boldsymbol{v}$ be the instantaneous velocity of a particle. The local average velocity of the particulate phase, $\boldsymbol{u}$, is of course:

$$\boldsymbol{u} = \langle \boldsymbol{v} \rangle \tag{2.1}$$

Let $\boldsymbol{c} = \boldsymbol{v} - \boldsymbol{u}$ be the instantaneous fluctuating velocity of the particle with respect to the local average velocity. $\boldsymbol{c}$ has of course zero average, $\langle \boldsymbol{c} \rangle = 0$. Since $\langle \boldsymbol{c} \cdot \boldsymbol{u} \rangle = 0$, the average total kinetic energy per unit mass of the particulate phase, $K_{TOT}$, is:

$$K_{TOT} = \langle v^2/2 \rangle = \langle u^2/2 \rangle + \langle c^2/2 \rangle \tag{2.2}$$

$\langle u^2/2 \rangle$ is the kinetic energy associated with the main motion of the particulate phase, $K$. In analogy with the classical Maxwell (1867) [but the credit should really go to Waterston, (1845)] statistical theory of rarefied gases, $\langle c^2/2 \rangle$ has been identified (to within a numerical factor) with the "granular temperature", $T$, so that:

$$K_{TOT} = K + T \tag{2.3}$$

Temperature is of course a thermodynamic concept, and therefore, although this had not been recognized until recently, the introduction of the concept of granular temperature is the first step towards the development of a large-scale thermodynamic theory of granular systems.

Indeed, Jenkins and Savage (1983) and Haff (1983) almost simultaneously introduced the idea of a "balance of granular temperature". A subtle conceptual point arises in this regard. In the Maxwellian theory of rarefied gases, the internal energy per unit mass, $U$, is proportional to temperature, the proportionality constant (the constant volume specific heat) being a (dimensional) universal constant. In classical gas dynamics, $U$ and $T$ do not have the same dimensions; should they have been assigned the same dimensions, the proportionality constant could easily have been set to unity by normalization, and the distinction between $U$ and $T$ would have become blurred. This is what has happened in the literature on granular systems, where $T$ has not been assigned its own dimensions, and hence $U$ and $T$ have essentially been identified with one another as if they were the same thing. It is however useful (and in one case to be discussed later it is crucially necessary) to keep the conceptual distinction between $U$ and $T$ in mind, even if the two quantities are assigned the same dimensions and indeed often the same value. The "balance of granular temperature" is then more clearly recognized as a balance of "granular energy", which clearly identifies it with the first law of large-scale thermodynamics.

Let $\boldsymbol{\sigma}$ be the particulate phase stress tensor. Regarding $\boldsymbol{\sigma}$ as the momentum flux tensor, as is usual in fluid mechanics, forces the convention that compressive stresses are positive. Let $\boldsymbol{D}$ be the particulate phase rate of strain tensor:

$$\boldsymbol{D} = [grad\boldsymbol{u} + grad\boldsymbol{u}^T]/2 \tag{2.4}$$

The net rate of work per unit volume done by internal stresses in the particulate phase (i.e., that part of the work which does not result in an increase of the kinetic energy of the main motion), $w$, is:

$$w = -\boldsymbol{\sigma} : \boldsymbol{D} \tag{2.5}$$

Let $q$ be the local average net influx of granular energy per unit volume. A non-zero $q$ may exist for two different reasons. First, there might be a direct influx $Q$ due, e.g., to vibration of the container; since one is trying to develop a theory of granular thermodynamics, it is useful to recognize $Q$ as the analog of the radiant heat influx in classical thermodynamics. Second, there may well be a conductive flux of granular energy, $\boldsymbol{q}$, related to the physically obvious possibility that granular energy may tend to move in the direction of decreasing granular temperature. Hence one has:

$$q = Q - \mathrm{div}\boldsymbol{q} \tag{2.6}$$

Finally, let $I$ be the rate of dissipation of granular energy, per unit volume, due to inelasticity of particle-particle collisions. In the Maxwellian theory, collisions between molecules are assumed to be perfectly elastic; however, such an assumption cannot be extended to granular systems. The first law of large scale thermodynamics (or the balance of granular energy) is written as ($D/Dt$ is the substantial derivative operator; in the classical inertial theories, $U = T$):

$$\phi DU/Dt + I = w + q \tag{2.7}$$

There is no analog of the term $I$ in classical thermodynamics, and this immediately shows that granular thermodynamics is of necessity a non-equilibrium theory. In the absence of a net influx of granular energy ($w+q=0$), the (positive) term I implies that $DU/Dt$ is negative, i.e., that granular energy steadily decreases: a granular system cannot stay thermalized without a continuous influx of energy, and hence its only "equilibrium" state is one where $T = 0$. Notice also that, since the only "equilibrium" state is $T = 0$, no argument of the "quasi-static process" kind is available - any non-trivial state of a granular system is not in any sense close to an equilibrium state. Another consequence of the presence of the term I is that in general granular energy is not equipartitioned (Maddox 1995, Ocone and Astarita 1995b).

In order to close the system of relevant equations, constitutive equations need to be written down for the particulate phase constitutive variables (or functions of state) $U, \boldsymbol{\sigma}, \boldsymbol{q}, I$, and for the interaction force between the interstitial fluid and particulate phases (Jackson 1986). These have been discussed in the literature, and their detailed form needs not concern us here: in order to develop a large scale thermodynamic theory, one only needs to establish the constitutive *class*, i.e., the list of independent state variables which determine the values of the constitutive properties (Truesdell 1984). This list, as deduced from the relevant literature, includes the following: solid volume fraction $\epsilon$, granular temperature $T$ and its gradient $\boldsymbol{n} = \text{grad}T$, rate of strain tensor $\boldsymbol{D}$. [Some of the published theories include grad$\epsilon$, in the list, but the thermodynamic analysis (Ocone and Astarita 1993) has shown that internal consistency requires grad$\epsilon$ to drop out of the list].

Having identified the constitutive class, one may proceed to write down the cornerstone of large scale statistical thermodynamics, i.e., the second law. Physically, the second law of granular thermodynamics aims at describing the "irreversibility" of the transfer of mechanical energy to the level of particle oscillations: when it is transferred, although still present in the form of kinetic energy, it is not entirely retrievable any more. In developing the formulation of the second law of granular thermodynamics, Ocone and Astarita (1993) have followed the philosophical approach of Truesdell (1984) and Coleman and Noll (1963), consistently with the fact that granular thermodynamics is a non-equilibrium theory where no analog of the vague "quasi-static process" idea can be used. Hence granular entropy $S$ (a concept without which no compressible flow theory can be developed) is regarded as a primitive concept which a), cannot be defined in terms of more fundamental concepts, and b), is only determined by the axioms laid down for it. The first axiom is that granular entropy is a function of state, and hence, by the principle of equipresence (Truesdell 1984), a constitutive function $s()$ exists which delivers the value of $S$:

$$S = s(\epsilon, T, \boldsymbol{n}, \boldsymbol{D}) \tag{2.8}$$

The second axiom is of course the second law of large scale thermodynamics, and here a subtle point arises in connection with the inelasticity of collisions. First consider the case where there is no net influx of entropy in the neighborhood of the point considered. In classical thermodynamics, one would for this case write the second law by requiring $DS/Dt$ to be non-negative. Here however the term I provides a new source of irreversibility, and hence the Ocone and Astarita (1993) approach is equivalent to writing down the second law for this case as (note that since $U$ and $T$ have the same dimensions, $S$ is dimensionless; also, $T$ is intrinsically non-negative, and, should it be zero, the quantity $I/T$, see Eq. 3.16 below, would be zero):

$$\phi DS/DT + I/T \geq 0 \tag{2.9}$$

i.e., for the case at hand granular entropy may increase, but not faster than $I/\phi T$.

For the general case, the net rate of influx of entropy per unit volume, $J$, is not zero, and thus the second law takes the form:

$$\phi DS/Dt + I/T \geq J \tag{2.10}$$

In analogy with the classical theory, $J$ is assumed to be made up of two parts. First, there may be a direct entropy influx per unit volume $Q/T$; second, there might be an unbalance of the conductive entropy flux $\boldsymbol{q}/T$:

$$J = Q/T - \text{div}(\boldsymbol{q}/T) = q/T + \boldsymbol{q} \cdot \boldsymbol{n}/T^2 \tag{2.11}$$

Let $Z = -\boldsymbol{q} \cdot \boldsymbol{n}/T$ be the "thermal dissipation rate", i.e., the rate of dissipation of granular energy due to its flowing towards regions of lower granular temperature. By substitution and rearrangement one obtains the final expression of the second law in its general form involving granular entropy:

$$\phi TDs/Dt + I + Z \geq q \tag{2.12}$$

It is as usual best to introduce the granular free energy $A = U - TS$, and to combine the first and second laws so as to obtain the following form involving granular free energy:

$$\phi[DA/Dt + SDT/Dt] \leq w + Z \tag{2.13}$$

**3. Constitutive properties.** Ocone and Astarita (1993) have followed the usual route (Coleman and Noll 1963) in order to deduce consequences of Eq. 2.13 for the constitutive class which we call "inertial" in the following. Before discussing their results, it is useful to decompose both $\boldsymbol{\sigma}$ and $\boldsymbol{D}$ into the sum of their isotropic and deviatoric parts ($\mathbf{1}$ is the identity tensor):

$$\boldsymbol{\sigma} = p\mathbf{1} + \boldsymbol{\sigma}^D \tag{3.1}$$

$$\boldsymbol{D} = (tr\boldsymbol{D}/3)\mathbf{1} + \boldsymbol{D}^D \tag{3.2}$$

$$tr\boldsymbol{\sigma}^D = tr\boldsymbol{D}^D = 0 \tag{3.3}$$

where $p$ is the particulate phase "pressure". This decomposition, in conjunction with the continuity equation for the particulate phase, allows us to decompose the net rate of work $w$ into a compressive part $w_c$ and a distortive part $w_D$:

$$w = (p/\epsilon)D\epsilon/Dt - \boldsymbol{\sigma}^D : \boldsymbol{D}^D = w_c + w_D \tag{3.4}$$

Let $a()$ be the constitutive function which delivers the value of $A$, $A = a(\epsilon, T, \boldsymbol{n}, \boldsymbol{D})$. The classical procedure for deriving conditions from the requirement that the second law should hold for all conceivable processes (Truesdell 1984) yields, in the case of Eq. 2.13 the following results (Ocone and Astarita 1993; subscripts indicate partial derivatives):

1). Free energy $A$ is in fact independent of $\boldsymbol{D}$ and $\boldsymbol{n}$:

$$A = a(\epsilon, T) \tag{3.5}$$

2). Entropy is minus the partial derivative with respect to $T$ of free energy, and is therefore also independent of $\boldsymbol{D}$ and $\boldsymbol{n}$:

$$S = -a_T = s(\epsilon, T) \tag{3.6}$$

3). It follows that also granular energy $U$ is independent of $\boldsymbol{D}$ and $\boldsymbol{n}$:

$$U = u(\epsilon, T) \tag{3.7}$$

4). If $A = U - TS$ and Eq. 3.7 is substituted into Eq. 3.6, one obtains:

$$s_T = u_T/T \tag{3.8}$$

5). The particulate phase pressure $p$ can be decomposed as the sum of a term independent of $D\epsilon/Dt$, $p^*$, and one which depends on it, $P$:

$$p = p^* + P \tag{3.9}$$

6). $p^*$ does only reversible work in any non-trivial compressible flow (i.e., for any nonzero $D\epsilon/Dt$), and it is given by:

$$p^* = \epsilon^2\phi a_\epsilon = F(\epsilon, T) \tag{3.10}$$

The existence of a finite value of $p^*$ insures that the net work done by particulate phase stresses is not entirely dissipative. In particular, the distortive rate of work is entirely dissipative, but at least part of the compressive rate of work is reversible.

7). $P$ does only irreversible work, i.e.:

$$PD\epsilon/Dt \geq 0 \tag{3.11}$$

This in turn implies, under mild assumptions of smoothness, that when $D\epsilon/Dt = 0$, $P = 0$.

8). As expected, the rate of dissipation due to inelasticity is non-negative:

$$I \geq 0 \tag{3.12}$$

9). If (as statistical theories suggest) $\boldsymbol{\sigma}^D$ is a linear function of $\boldsymbol{D}^D$, $\boldsymbol{\sigma}^D = 2\mu\boldsymbol{D}^D$, and $P$ is a linear function of $D\epsilon/Dt$, $P = \beta D\epsilon/Dt$, the viscosity coefficients are required to be non-negative:

$$\mu \geq 0; \quad \beta \geq 0 \tag{3.13}$$

It is interesting to note that, in the classical Maxwellian theory, $\beta = 0$ identically: pure compressions and expansions of a Maxwellian gas are reversible (Truesdell 1984). This does not carry over to granular systems.

10). Finally, if the constitutive equation for $\boldsymbol{q}$ is written in the form of Fourier's law, $\boldsymbol{q} = -k\boldsymbol{n}$, the thermal conductivity is required to be non-negative:

$$k \geq 0 \tag{3.14}$$

Points $1 - 10$ are of general validity, i.e., they hold in both the inertial and the viscous limit (to be discussed below) and in any combination thereof. The decomposition of pressure in Eq. 3.9 is of particular relevance, since $p^*$ is recognized as an analog of the "equilibrium" pressure: the actual pressure $p$ is equal to $p^*$ when the granular material neither expands nor contracts, i.e., when $D\epsilon/Dt = 0$. This immediately suggests to define granular enthalpy $H$ as $U + p^*/\phi$, and granular free enthalpy $G$ as $A + p^*/\phi$.

The inertial limit is the case most usually considered in the literature, where parameters entering the constitutive equations for the particulate phase are $d_p, \phi_p$, the coefficient of restitution $\alpha$ (with $\alpha = 1$ corresponding to elastic collisions), in addition to the state variables: the important point to stress is that no property of the interstitial fluid enters the constitutive equations for the particulate phase. This of course classifies theories of dry granular flow, where no interstitial fluid is present [such as in the analysis of the dynamics of the Saturn rings (Borderies et al. 1985, Goldreich and Tremaine 1978)], as inertial limit ones.

In the inertial limit, the form of the constitutive equations can, in its essential part, be deduced simply from dimensional analysis, except for the dependence on $\epsilon$ which dimensional analysis leaves of course undefined. In particular, the "conservative" pressure $p^*$, and the dissipation rate $I$, are obviously given by (with $f()$ and $h()$ being defined as having non-negative values):

$$p^* = f(\epsilon)\phi_p T \tag{3.15}$$

$$I = (1-\alpha)h(\epsilon)\phi_P T^{3/2}/d_p \tag{3.16}$$

An important point should be stressed here. A general theorem of classical thermodynamics (Astarita and Sarti 1975) can be extended to large scale thermodynamics. For the case at hand, the theorem may be worded as follows:

"Materials with 'entropic elasticity' are materials for which energy $U$ is a unique function of temperature $T$. For such materials, the 'reversible' pressure $p^*$ is a linear function of $T$".

In the classical inertial limit type of theories, not only $U$ is regarded as a unique function of $T$ [$U = u(T)$], it is in fact identified with it; hence inertial-limit granular materials are endowed with entropic elasticity (thus falling, in this sense, in the same category as ideal gases and ideal rubbers), and indeed Eq. 3.15 shows that $p^*$ is linear in $T$. The following result holds for materials with entropic elasticity:

11). As can be seen by substituting $U = u(T)$ into Eq. 3.8 and differentiating the result with respect to $\epsilon$, the second derivative of granular entropy with respect to $\epsilon$ and $T$ is zero, and hence $S$ is the sum of a function of only $T$ and a function of only $\epsilon$:

$$S = s_1(T) + s_2(\epsilon) \tag{3.17}$$

(Notice that this trivially guarantees that $s_{\epsilon T} = s_{T\epsilon}$).

The result in Eq. 3.17, and the preceding results, make it possible to obtain an explicit equation for the differential change of granular entropy, $\delta S$, corresponding to arbitrary differential changes of $T$ and $\epsilon$ [with $u'$ the ordinary derivative of $u()$]:

$$\delta S = u'(T)\delta T/T + f(\epsilon)\delta\epsilon/\epsilon^2 \tag{3.18}$$

which can of course be integrated to obtain an explicit equation for the granular entropy (given the fact that $u'(T) = 1$; $S^*$ is the granular entropy corresponding to some reference state $\epsilon^*$, $T^*$; $w$ is the dummy variable):

$$S - S^* = ln(T/T^*) - \int_{\epsilon^*}^{\epsilon} f(w)dw/w^2 \tag{3.19}$$

Although the body of this paper is restricted to the inertial limit, it is conceptually illuminating to analyze the viscous limit as well. The viscous limit case has been introduced by McTigue and Jenkins (1992). The case considered is the one where the interstitial fluid viscosity is large enough that, when two particles approach each other, lubrication theory type of forces related to the squeezing flow between the two particles become important. The detailed analysis of the viscous limit needs not concern us here; following Ocone and Astarita (1995c), we only present the essential granular thermodynamics results for this case.

In the viscous limit, the interstitial fluid viscosity $\mu_F$ appears in the list of parameters, but the particle density does not. Hence Eq. 3.15 takes, as can be seen by a trivial dimensional argument, the following form (we indicate with a caram constitutive functions in the viscous limit):

$$p^* = f^\wedge(\epsilon)\mu_F\sqrt{T}/d_P \tag{3.20}$$

Since $p^*$ is not linear in $T$, the converse implication of the general theorem presented above implies that *in the viscous limit $U$ is not a unique function of $T$*. This clearly shows that the conceptual distinction between $U$ and $T$ is crucially important as soon as the interstitial fluid viscosity plays some role in determining the stress tensor in the particulate phase. Ocone and Astarita (1995c) have presented a complete thermodynamic analysis of both the viscous limit and the case intermediate between the inertial and viscous limits. The essential results are as follows:

a). Let $R = \phi_p d_p\sqrt{T}/\mu_F$ be a "Reynolds number". $\mu_F/d_p\phi_p$ is an intrinsic velocity scale. The inertial limit corresponds to $R >> 1$ or $\sqrt{T} >> \mu_F/d_p\phi_p$, the viscous limit to $R << 1$, $\sqrt{T} << \mu_F/d_p\phi_p$. For the general case, $p^*$ is given by:

$$p^* = \phi_P f(\epsilon)T + \mu_F f^\wedge(\epsilon)\sqrt{T}/d_P \tag{3.21}$$

b). If $\sqrt{T}$ scales with the particles terminal velocity, the condition $R << 1$ can be expressed as $Ar << 1$, where $Ar$ is the Archimedes number. If $\sqrt{T}$ scales with some externally imposed velocity scale, it can be expressed as $Re_E << 1$, where $Re_E$ is a Reynolds number based on the external velocity scale and the particle diameter. In both cases, $T$ itself does not appear explicitly in the condition.

c). Granular Energy $U$ is given by:

$$U = T + u^\wedge(\epsilon)\mu_F\sqrt{T}/d_p\phi_p \tag{3.22}$$

d). The condition that $s_{\epsilon T} = s_{T\epsilon}$ is not trivially satisfied, and indeed it implies that functions $u^\wedge()$ and $f^\wedge()$ are related to each other by (with $u^{\wedge\prime}$ the ordinary derivative of $u^\wedge$):

$$f^\wedge(\epsilon)/\epsilon^2 = 2u^{\wedge\prime}(\epsilon) \tag{3.23}$$

e). The classical Maxwell relations hold in the general case (which of course includes the inertial and viscous limits as special asymptotic cases):

$$\delta U = T\delta S + p^*\delta\phi/\phi^2 \tag{3.24}$$
$$\delta H = \delta p^*/\phi + T\delta S \tag{3.25}$$
$$\delta A = -S\delta T + p^*\delta\phi/\phi^2 \tag{3.26}$$
$$\delta G = \delta p^*/\phi - S\delta T \tag{3.27}$$

f). Given arbitrary variations $\delta\epsilon, \delta T$, the corresponding $\delta S$ is given by:

$$\delta S = -[f(\epsilon)/\epsilon^2 + u^{\wedge\prime}(\epsilon)/R]\delta\epsilon + [1/T + u^{\wedge}(\epsilon)/2RT]\delta T \tag{3.28}$$

**4. Elementary compressible flow problems.** The most elementary problem of compressible flow in classical gas dynamics is the determination of the speed of sound; this problem was tackled as early as 1678 by Newton, and the correct solution was obtained by Laplace in 1816. Of course neither Newton nor Laplace had any thermodynamic theory available to use as a basis for their estimates of the speed of sound.

The corresponding problem for granular materials was first analyzed by Savage (1988), who again did not have a granular thermodynamic theory available, and yet he arrived at the correct result just the way Laplace did for the classical problem. The complete, thermodynamics-based analysis was given by Ocone and Astarita (1994). The problem to be considered is the speed of propagation of an imposed infinitesimal granular pressure discontinuity $\delta p$. In the classical problem, the infinitesimal pressure wave propagates through a quiescent gas at some temperature $T$. This cannot blindly be extended to the corresponding granular problem, since a quiescent granular material cannot be thermalized, and hence $T = 0$ in such a system. One therefore needs to consider the superposition of of infinitesimal pressure wave on some preexisting condition which keeps the system at some steady granular temperature $T$; this might be, e.g., a shear flow or an externally vibrated system. [It is important to realize that the speed of propagation of an infinitesimal particulate phase pressure wave has nothing to do with the actual speed of sound in a granular system; real acoustic waves in granular systems have also been analyzed (Atkinson and Kytomaa 1992)].

One only needs to write down the mass and momentum balances to obtain the result that the square of the speed of propagation $V_s$ is given by (the subscript $S$ identifies the "pseudo-sonic" speed):

$$V_s^2 = \delta p^*/\delta\phi = E/\phi \tag{4.1}$$

where $\delta p^*$ and $\delta\phi$ are the (infinitesimal) jumps of pressure and density across the discontinuity (the result will be derived as a special case of a finite discontinuity below). Eq. 4.1 also defines the "particulate phase elastic modulus" $E$, which has been considered in a number of analyses of the stability of fluidized beds, where the limit condition for stability is the one where kinematic waves travel at the same speed as the dynamic waves governed by Eq. 4.1 (Batchelor 1988; Ding and Gidaspow 1990; Foscolo and Gibilaro 1987; Foscolo et al. 1995; Jean and Fan 1992; Mutsers and Rietema 1977; Rosensweig 1979). The problem is that, since $p^*$ is not a unique function of $\phi$ (or, equivalently, the value of $E$ is not known a priori), Eq. 4.1 does not determine unequivocally the value of $V_s^2$.

Leonardo da Vinci (1500) was the first one to recognize that sound propagates without any actual downstream motion of the supporting medium; he compared the propagation of sound to the waves that wind sends down a field of grain, where grain itself does not of course move down the field. Newton (1678) got as far as Eq. 4.1 in his analysis of the speed of sound. He then proceeded to assume that $\delta p^*/\delta\phi$ was to be the isothermal inverse compressibility of an ideal gas, and he got a result which was in error by about 30% in the estimate of the speed of sound in air at atmospheric conditions. Laplace (1816) reasoned that as the wave moves through the gas, compression takes place very quickly (nowadays, we would say adiabatically), and, by substituting the experimentally known $p^*(\phi)$ relationship for very fast compression he obtained the correct result. The Savage (1988) analysis was essentially based on an argument which is conceptually similar (at least vaguely so) to the Laplace one. Savage used classical equations available for granular flow theory; he then assumed that viscous effects and granular energy fluxes effects are negligible, thus essentially removing all causes of irreversibility; finally, he assumed that wave propagation was essentially "granular adiabatic", and he therefore proceeded to obtain the correct result. (In the granular thermodynamic analysis discussed below, a reversible, granular adiabatic process is of course granular isoentropic).

The physically convincing analysis was presented by Ocone and Astarita (1994). In addition to the balances of mass and momentum, one needs to write the balance of energy, and this yields the result that the derivative $\delta p^*/\delta\phi$ in Eq. 4.1 has to be intended as the inverse *isoentropic* compressibility. One then uses Eq. 3.28 (or, in the inertial limit, Eq. 3.18) to establish the relationship between $\delta T$ and $\delta\epsilon$ needed to insure that $\delta S = 0$ to obtain, in the inertial limit, the same result as was originally obtained by Savage ($f'$ being the ordinary derivative of the function $f$):

$$V_s^2 = T[f'(\epsilon) + f^2(\epsilon)/\epsilon^2] \tag{4.2}$$

The Newton (wrong) analysis corresponds to keeping only the $f'(\epsilon)$ (isothermal) term in Eq. 4.2. Given reasonable forms for $f(\epsilon)$, Ocone and Astarita (1994) show that the term $f^2(\epsilon)/\epsilon^2$ is at least as large as the $f'(\epsilon)$ one.

Ocone and Astarita (1995a) have extended the analysis to the propagation of finite discontinuities; the latter of course include infinitesimal ones as a special case. Let [b] represent the difference between the values of any quantity $b$ behind ($b_R$) and in front ($b_L$) of the shock which is regarded as stationary. Let $u$ be the velocity of the particulate phase [$u$ has been used as the modulus of the average velocity of the particulate phase, which for the case at hand coincides with the axial velocity], and $W = \phi u$ its mass flowrate per unit cross-sectional area. The mass balance yields the obvious result:

$$[\phi u] = [W] = 0 \tag{4.3}$$

The momentum balance (on both sides of the shock $D\epsilon/Dt = 0$ and hence $p = p^*$) yields:

$$[p^*] = -W^2[1/\phi] \tag{4.4}$$

Eq.s 4.3-4.4 can be combined to yield:

$$(\phi_1/\phi_2)V^2 = [p^*]/[\phi] \tag{4.5}$$

This of course reduces to Eq. 4.1 when the discontinuity is an infinitesimal one. Finally, the balance of energy yields the result that the jump of $H + u^2/2$ (the so called "total enthalpy") is zero, which can be reduced to:

$$[H] = [p^*](1/\phi_R + 1/\phi_L)/2 \tag{4.6}$$

In the case of an infinitesimal discontinuity, Eq. 4.6 reduces to $\delta H = \delta p^*/\phi$, and since the Maxwell relation in Eq. 3.25 holds true, $\delta S = 0$ and the classical result is obtained. For finite shocks, combination of Eq's 4.3-4.6 yields the equivalent of a classical Hugoniot (1887, 1888) curve; detonations ($[p^*] > 0$) travel at "supersonic" speeds, deflagrations ($[p^*] < 0$) travel at "subsonic" speed.

**5. Moderately complex problems of compressible flow.** The two problems considered in the previous section are comparatively trivial; the propagation of infinitesimal discontinuities is an essentially linear problem, and in the analysis of the propagation of finite shocks their *existence* is assumed at the start, so that the classical elementary analysis is easily extended from gas dynamics to granular flow.

As soon as one considers even moderately more complex problems, the situation is quite different. Consider for instance the case of a compression or a rarefaction wave. In classical gas dynamics, compression waves have a tendency to reinforce themselves into shocks, while rarefaction waves tend to smooth themselves out (which also implies that rarefaction shocks, or deflagrations, are intrinsically unstable). The approach to the analysis of waves (rather than shocks) in granular systems looks deceivingly similar to what is commonly done in classical gas dynamics. However, one soon identifies an entirely new problem, which is briefly discussed below.

Consider the simplest possible case where the granular system is kept at a steady thermalized state by vibration of the container, so that $Q$ and $I$ balance each other and thus:

$$T = [Qd_p/(1-\alpha)h(\epsilon)\phi_p]^{2/3} \tag{5.1}$$

(Notice that, when $\alpha \rightarrow 1$, $T \rightarrow \infty$. This is related to the fact that, if $Q$ cannot be balanced by $I$ because collisions are elastic, $T$ diverges because there is no way of disposing of the energy influx $Q$ - contrary to classical gas dynamics, the system cannot be maintained at constant granular temperature by heat removal through the constraining walls).

Eq. 5.1 imposes a relationship between possible couples of values of $\epsilon$ and $T$ which can be maintained at steady state. Condition 5.1 is guaranteed to be satisfied before the shock hits the system, i.e., by $T_L$ and $\epsilon_L$. As the (assumed) shock moves through, the values of $T_R$ and $\epsilon_R$ resulting behind it, which can be calculated from Eq's 4.3-4.6, will not satisfy Eq. 5.1. Indeed, consider for instance a compression shock, for which, according to Eq's 4.3-4.6, $T_R \geq T_L$ and $\epsilon_R \geq \epsilon_L$. Eq. 5.1 is certain not to be satisfied [except in the trivial case], since it gives $T$ as a decreasing function of $\epsilon$. Hence the shock, if indeed it can be maintained at all, is of necessity followed by a wake where thermal relaxation takes place, until (presumably) far downstream Eq. 5.1 is satisfied again and a new steady state is reached. (Such a thermal relaxation is governed by a nonlinear first order differential equation which is trivially deduced from the energy balance). There is no equivalent of such a thermal relaxation phenomenon in classical gas dynamics, where any couple of values of temperature and density can be maintained indefinitely. This simple argument shows what a crucial effect the dissipation term I has on the fundamental structure of granular compressible flow.

The difficulties involved are best appreciated by trying to follow classical gas dynamics in the formulation of the general one-dimensional granular compressible flow problem. Consider a smooth variation of, say, $\epsilon$, initially imposed on a one-dimensional flow field of the particulate phase such as plug flow down a vibrated conduit. Reinforcement of compression waves is simply due to the fact that the speed of sound increases with increasing temperature, so that successive infinitesimal compression disturbances propagate at an increasing speed. In a particulate phase, the speed of pseudo-sound also increases with increasing pseudo-temperature, but the latter undergoes the relaxation phenomenon discussed before, and this, as will be seen, makes a lot of difference.

Let $x$ be the spatial coordinate in the flow direction, $m$ the difference $Q - I$, and $t$ the time. Based on the usual approximations which are standard in gas dynamics (essentially, that viscosity and thermal conductivity are set to zero), and restricting attention to the inertial limit, the balance equations to be written are (we do not distinguish between $p$ and $p^*$ in the following, for obvious reasons):

mass balance:

$$\partial\phi/\partial t + u\partial\phi/\partial x + \phi\partial u/\partial x = 0 \tag{5.2}$$

momentum balance:

$$\partial u/\partial t + u\partial u/\partial x + (1/\phi)\partial p/\partial x = 0 \tag{5.3}$$

energy balance:

$$\partial U/\partial t + (p/\phi)\partial u/\partial x + u\partial U/\partial x = m/\phi \tag{5.4}$$

The culprit is the energy balance, because of the appearance of the term $m$ which makes the system of Eq.s 5.2-5.4 non-homogeneous.

We now try to develop the same compact method of solution which is standard in gas dynamics. Pressure $p$ is a unique function of $\phi$ and $U$, and the following two quantities can be defined:

$$M = (\partial p/\partial \phi)_U \tag{5.5}$$

$$X = (\partial p/\partial U)_\phi \tag{5.6}$$

We also define the following two vectors and the following matrix:

$$\boldsymbol{b} = (\phi, u, U) \tag{5.7}$$

$$\boldsymbol{p} = (0, 0, m/\phi) \tag{5.8}$$

$$\boldsymbol{A} = \begin{vmatrix} u & \phi & 0 \\ M/\phi & u & X/\phi \\ 0 & p/\phi & u \end{vmatrix} \tag{5.9}$$

so that the system of equations 5.2-5.4 reduces to:

$$\partial \boldsymbol{b}/\partial t + \boldsymbol{A}\partial \boldsymbol{b}/\partial x = \boldsymbol{p} \tag{5.10}$$

Contrary to the case of gas-dynamics, where $\boldsymbol{p} = 0$, the quasi-linear Eq. 5.10 is non-homogeneous. However, just as in the case of gas dynamics, the three eigenvalues of $\boldsymbol{A}$, which define the characteristic curves of the hyperbolic system 5.10, are $u, u - V_s$, and $u + V_s$.

The system 5.10 represents a typical hyperbolic problem (Rhee et al. 1989, Lax 1973, Courant and Hilbert 1953), and the classical approach is to diagonalize it. Suppose one can find the three Riemann invariants of 5.10, which would constitute a 3-dimensional vector $\boldsymbol{y}$. In that case, the system 5.10 reduces to:

$$\partial \boldsymbol{y}/\partial t + \boldsymbol{B}\partial \boldsymbol{y}/\partial x = \boldsymbol{C}\boldsymbol{p} \tag{5.11}$$

Here matrix $\boldsymbol{B}$ is a diagonal matrix, so that the three components of Eq. 5.11 are uncoupled. The elements of $\boldsymbol{B}$ are the eigenvalues of $\boldsymbol{A}$:

$$\boldsymbol{B} = \boldsymbol{C}\boldsymbol{A}\boldsymbol{C}^{-1} \tag{5.12}$$

and $C = \partial \boldsymbol{y}/\partial \boldsymbol{b}$ is a matrix whose lines are the left eigenvectors of $\boldsymbol{A}$:

$$(5.13) \qquad C = \begin{vmatrix} p/\phi^2 & 0 & -1 \\ M/X & -V_s\phi/X & 1 \\ M/X & V_s\phi/X & 1 \end{vmatrix}$$

The Riemann invariants can in fact be found if each row of $C$ derives from a pseudo-potential, say if $\boldsymbol{y}$ is expressible as $r(\boldsymbol{b})\text{grad}z(\boldsymbol{b})$, with $r()$ and $z()$ being scalar-valued functions. This in turn implies that the following equation holds for $i = 1, 2, 3$ (Morese and Feshbach 1953):

$$(5.14) \qquad \text{curl}(\partial y_i/\partial \boldsymbol{b}) \cdot (\partial y_i/\partial \boldsymbol{b}) = 0$$

Since the relation between $p$ and $T$ is linear (which is true for ideal gases as well as for granular materials), $\partial y_1/\partial \boldsymbol{b}$ admits the integrating factor $1/T$ and hence it can be directly integrated. Eq. 5.14 is satisfied since $\partial y_1/\partial \boldsymbol{b}$ is solenoidal. It is important to realize that $y_1$ is in fact the (granular) entropy, and that the equation for it is:

$$(5.15) \qquad DS/Dt = \partial S/\partial t + u\partial S/\partial x = m/\phi T$$

In gas dynamics, $m = 0$, and hence Eq. 5.15 implies that in the absence of shocks, if the initial conditions are isoentropic, they will stay isoentropic at all times. This in turn implies that Eq.5.4 can be discarded and substituted by the condition of isoentropicity. Thus $p$ reduces to a unique function of $\phi$, $dp/d\phi = V_s^2$, and the problem formulation reduces to only two equations. Eq. 5.14 is satisfied by the reduced set of equations, and hence the Riemann invariants are established. With this, the system reduces to two ordinary differential equations along the characteristic curves, and the solution can be obtained along the classical lines.

However, Eq. 5.15 shows that, in the case of granular flow, even if the initial conditions are iso-pseudo-entropic, they will not stay so at later times - and one cannot follow the shortcut of classical gas dynamics. This is related to the fact that the thermal relaxation phenomenon discussed before is not isoentropic; and relaxation takes place only when $m$ is nonzero. In the system of Equations 5.2-5.4, $m$ is a lower order term, and yet its presence gives an entirely new structure to the problem.

At this stage, we have concluded that the classical approach for the problem considered cannot be followed in the case of granular flow. Formally, there is also the difficulty that, since the Riemann invariants cannot be found, there is no guarantee of existence and uniqueness of the solution; this however we believe is not a problem. Of course, Eq.s 5.2-5.4 could be solved numerically for any assigned initial condition, but the numerical techniques involved are known to result, even in the case of ideal gases, in rather sever numerical problems (Richtmeyer and Morton 1967). However, we are interested first in establishing the structure of the analysis of compressible flow of granular materials, and we analyze another problem in the following.

Solution of the entire set of three equations is done in classical gas dynamics when on the one side a shock is regarded as already existing, and on the other side wants to determine the "thickness" of, say, a compression wave. The thickness will be finite if one includes the effects of a nonzero bulk viscosity, $\beta$, and thermal conductivity, $k$. Granular materials have indeed nonzero $\beta$ and $k$ (Jenkins and Savage 1983, Haff 1983, Jackson 1986). Inclusion of bulk viscosity and thermal conductivity makes the problem parabolic (in the unsteady state formulation), or elliptic (in the steady state one), so that the question of possible existence of Riemann invariants disappears entirely.

Consider plug flow of a granular material, kept thermalized by vibration, down a constant section conduit. Consider a wave which is moving in the negative $x$ direction at some speed $V$; the wave can be regarded as stationary if one subtracts $V$ from all velocities. The resulting standing wave problem is elliptic, and it bears of course a strong analogy with the standing wave problem in classical gas dynamics.

The mass, momentum and energy balances are:

$$d(\phi u)/dx = 0 \tag{5.16}$$

$$\phi u \, du/dx = -dp/dx + d(\beta du/dx)/dx \tag{5.17}$$

$$\phi u \, dU/dx = -p \, du/dx + \beta(du/dx)^2 + d(kdT/dx)/dx + m \tag{5.18}$$

The second term on the right of Eq. 5.18 represents the generation of pseudo-energy associated with stresses due to the bulk viscosity $\beta$. As usual, it is useful to multiply the momentum balance by $u$ and sum the result with the energy balance, so as to obtain:

$$\phi u d(U + u^2/2)/dx = -d(pu)/dx + d(u\beta du/dx + kdT/dx)/dx + m \tag{5.19}$$

Since $d(pu)/dx = \phi u d(p/\phi)/dx$, one can use also the balance of mass to transform Eq. 5.19 to the following useful form:

$$\phi u d(U + u^2/2 + p/\phi)/dx = d(u\beta du/dx + kdT/dx)/dx + m \tag{5.20}$$

The quantity $U + u^2/2 + p/\phi$ is what is usually called the "total enthalpy" in gas dynamics, i.e., the sum of the actual enthalpy $H = U + p/\phi$ and the kinetic energy per unit mass.

In principle one would need to determine the distributions of $u, p, \phi$, and $U = T$ within the wave. Of these, only three are independent, since $p$ is a unique function of $\phi$ and $U$. The three distributions can be obtained, in principle, by integrating Eq.s 5.16, 5.17 and 5.20. If Eq. 5.20 is integrated over the whole thickness of the wave, the terms including $\beta$ and $k$ integrate to zero.

The balance of energy, however, again poses a problem. Let $N$ be defined as the total unbalance between pseudo-energy supply and dissipation within the wave:

$$N = \int_L^R m dx \tag{5.21}$$

Integrating Eq. 5.20 between $L$ and $R$ yields:

$$T_L + p_L/\phi_L + u_L^2/2 + N = T_R + p_R/\phi_R + u_R^2/2 \tag{5.22}$$

which implies that the total pseudo-enthalpy on the right is different from what it is on the left (the sign of $N$ cannot be established a priori).

In classical gas dynamics, $N = 0$, and hence the total enthalpy is the same on the two sides; from this one obtains the classical Hugoniot conditions relating temperature, velocity and density on the two sides of a shock. The Hugoniot conditions have originally been derived for an existing shock (not a finite-thickness standing wave), and in their derivation bulk viscosity and thermal conductivity are set at zero. However, the equations apply in spite of the fact that one is considering a wave of finite thickness and not a shock. The bulk viscosity and the conductivity are non-zero, and their values determine the thickness of the wave, as well as the distributions of $T$, $u$ and $p$ within it, but they do not influence the relationship between the values of these quantities on the right and on the left.

In the case of granular materials, Eq. 5.22 shows that the standing wave equations do not result in the same left and right conditions as those of a shock; this of course implies that the very possibility of the existence of finite shocks cannot be taken for granted. The mathematical status of the problem is as follows. The analysis of an existing shock is in fact the "outer" solution of a classical singular perturbation problem (Nayfeh 1973), and in this sense the standing wave is an inner boundary layer. In gas dynamics, the inner boundary layer equations integrate to the shock ones, indicating that the outer solution is a proper asymptote. This is not the case for granular systems, for which the asymptotic status of the shock equations is unclear.

We are currently investigating ways of solving numerically the standing wave equations for granular materials, but again, numerical techniques which have been developed in classical gas dynamics do not carry over in any easily identified way.

**6. Conclusions.** The background of a large scale theory of granular thermodynamics, which is needed in order to develop a theory of compressible granular flow, has recently been made available. This makes it possible to set up in a rational way problems of compressible granular flow. The most elementary problems can be solved following traditional lines of analysis developed for classical gas dynamics. However, as soon as one considers even only moderately complex problems, the inelasticity of particle-particle collisions (of which there is no analog in classical gas dynamics) entirely changes the mathematical structure of the problems, and classical gas dynamics methods cannot be extended to compressible granular flow. The difficulties have been identified, but at this stage there is no way of overcoming them.

**7. Notation.** Note: Thermodynamic quantities do not always have the same dimensions in classical and granular thermodynamics. When this is the case, the classical dimensions are given in parentheses.

Some vectors and matrices are ordered arrays of quantities having different dimensions, and hence no dimensions are given for them.

Dimensionless quantities have dimensions indicated with -

| | |
|---|---|
| $a(\ )$ | Function delivering granular free energy, $m^2/s^2$ |
| $A$ | Free energy, $m^2/s^2$ |
| $\boldsymbol{A}$ | Matrix defined in Eq. 5.9 |
| $b$ | Any quantity |
| $\boldsymbol{b}$ | $= (\phi, u, U)$ |
| $\boldsymbol{B}$ | Diagonal matrix of the eigenvectors of $\boldsymbol{A}$ |
| $c$ | Fluctuation velocity, $m/s$ |
| $\boldsymbol{C}$ | $= \partial \boldsymbol{y}/\partial \boldsymbol{b}$ |
| $d_p$ | Particle diameter, $m$ |
| $\boldsymbol{D}$ | Rate of strain, $s^{-1}$ |
| $\boldsymbol{D}^d$ | Deviatoric part of $\boldsymbol{D}, s^{-1}$ |
| $E$ | Elastic modulus, $Kg/m, s^2$ |
| $f(\ )$ | Function defined in Eq. 3.15,- |
| $f^\wedge(\ )$ | Function defined in Eq. 3.20,- |
| $F(\ )$ | Function defined in Eq. 3.10, $Kg/m, s^2$ |
| $G$ | Free enthalpy, $= H - TS, m^2/s^2$ |
| $h(\ )$ | Function defined in Eq. 3.16, - |
| $H$ | Enthalpy, $= U - p^*/\phi, m^2/s^2$ |
| $i$ | Index, $i = 1, 2, 3$, - |
| $I$ | Rate of dissipation due to inelasticity, $kg/m, s^3$ |
| $J$ | Rate of entropy influx, $1/m^3, s(1/K, m, s^3)$ |
| $k$ | Thermal conductivity, $kg/m, sec(Kg/s^2)$ |
| $K$ | Kinetic energy of mean motion, $m^2/s^2$ |
| $K_{\mathrm{TOT}}$ | Total kinetic energy, $m^2/s^2$ |
| $\ell$ | Scale of averaging, $m$ |
| $L$ | Scale of apparatus, $m$ |
| $m$ | $= Q - I, Kg/m, s^3$ |
| $M$ | $= (\partial p/\partial \phi)_U, m^2/s^2$ |
| $\boldsymbol{n}$ | temperature gradient, $m/s^2(K/m)$ |

| | |
|---|---|
| $N$ | Total collision dissipation within the wave, $Kg/s^3$ |
| $p$ | Particulate phase pressure, $Kg/m,s^2$ |
| $p^*$ | Particulate phase reversible pressure, $Kg/m,s^2$ |
| $\boldsymbol{p}$ | $= (0,0,m/\phi)$ |
| $P$ | Irreversible pressure, $Kg/m,s^2$ |
| $q$ | Local influx of energy, $Kg/m,s^3$ |
| $\boldsymbol{q}$ | Energy flux, $Kg/s^3$ |
| $Q$ | Direct energy influx, $Kg/m,s^3$ |
| $R$ | $= d_p\phi_p\sqrt{T}/\mu_F, -$ |
| $Re_E$ | Reynolds number based on external velocity scale, - |
| $r(\boldsymbol{b})$ | Scalar-valued function |
| $s(\ )$ | Function delivering entropy, - $(m^2/K,s^2)$ |
| $s_1(T)$ | Temperature part of entropy, - $(m^2/K,s^2)$ |
| $s_2(\epsilon)$ | Density part of entropy, - $(m^2/K,s^2)$ |
| $S$ | Entropy, - $(m^2/K,s^2)$ |
| $t$ | Time, $s$ |
| $T$ | Temperature, $m^2/s^2(K)$ |
| $u(\ )$ | Function delivering energy, $m^2/s^2$ |
| $u^\wedge(\ )$ | Function defined in Eq. 3.22, - |
| $u$ | Mean motion velocity modulus, $m/s$ |
| $\boldsymbol{u}$ | Local average velocity vector, $m/s$ |
| $U$ | Energy, $m^2/s^2$ |
| $\boldsymbol{v}$ | Particle velocity vector, $m/s$ |
| $V$ | Wave speed, $m/s$ |
| $V_s$ | Speed of pseudo-sound, $m/s$ |
| $w$ | Local net rate of work, $Kg/m,s^3$ |
| $w_c$ | Compressive rate of work, $Kg/m,s^3$ |
| $w_D$ | Distortive rate of work, $Kg/m,s^3$ |
| $W$ | Mass flowrate, $Kg/s$ |
| $x$ | Axial position, $m$ |
| $X$ | $= (\partial p/\partial U)_\phi, Kg/m^3$ |
| $\boldsymbol{y}$ | Riemann invariants in Eq. 5.10 |
| $z(\boldsymbol{b})$ | Scalar-valued function |
| $Z$ | Thermal dissipation rate, $Kg/m,s^3$ |
| $\alpha$ | Coefficient of restitution, - |
| $\beta$ | Bulk viscosity, Kg/m,s |
| $\epsilon$ | Solid volume fraction, - |
| $\mu$ | Particulate phase shear viscosity, Kg/m,s |
| $\mu_F$ | Fluid viscosity, Kg/m,s |
| $\phi$ | Particle phase local density, $Kg/m^3$ |
| $\phi_P$ | Particle density, $Kg/m^3$ |
| $\boldsymbol{\sigma}$ | Particulate phase stress tensor, $Kg/m,s^2$ |
| $\boldsymbol{\sigma}^D$ | Deviatoric part of $\boldsymbol{\sigma}, Kg/m,s^2$ |
| $\mathbf{1}$ | Unit tensor, - |

Operators:

| | |
|---|---|
| $D/Dt$ | Substantial derivative, $s^{-1}$ |
| grad | Gradient, $m^{-1}$ |
| tr | Trace of a tensor, - |
| $\delta$ | Infinitesimal variation, - |
| [ ] | Difference of values on two sides of shock, - |
| $\langle\ \rangle$ | Local average, - |
| : | Scalar product of two tensors, - |

Superscripts

| | |
|---|---|
| $*$ | In reference state |
| $\wedge$ | In viscous limit |
| $T$ | Transpose, - |

Subscripts

| | |
|---|---|
| $L$ | On the left of shock or wave |
| $R$ | On the right of shock or wave |

## REFERENCES


[1] Anderson, T.B., and R. Jackson, *A Fluid Mechanical Description of Fluidized Beds*, Ind. Eng. Chem. Fund., 6, 527–539 (1967).

[2] Astarita, G., *Thermodynamics. An Advanced Textbook for Chemical Engineers*, Plenum, New York, 1989.

[3] Astarita, G., and Sarti, G.C., *Thermomechanics of Compressible Materials with Entropic Elasticity*, in *Theoretical Rheology*, J.F. Hutton, J.R.A. Pearson and K. Walters Eds., Applied Science Publ., Barking, 1975.

[4] Astarita, G., and R. Ocone, *Large-Scale Statistical Thermodynamics and Wave Propagation in Granular Flow*, I& EC Research, 33, 2280–2287 (1994).

[5] Atkinson, C.M., and H.K. Kytomaa, *Acoustic Wave Speed and Attenuation in Suspensions*, Intl. J. Multiphase Flow, 15, 577–592 (1992).

[6] Bagnold, R.A., *Experiments on a Gravity-Free Dispersion of Large Solid Particles in a Newtonian Fluid under Shear*, Proc. Roy. Soc. London, A225, 49–63 (1954).

[7] Batchelor, G.K., *A New Theory for the Instability of a Uniform Fluidized Bed*, J. Fluid Mech., 193, 75–110 (1988).

[8] Borderies, N.P. Goldreich, and S.A. Tremaine, *A Granular Flow Model for Dense Planetary Rings*, Icarus, 63, 406–420 (1985).

[9] Bowen, R.M., *Continuum Physics*, A.C. Eringen Ed., Academic Press, vol. II, New York, 1971.

[10] Coleman, B.D., and W. Noll, *The Thermodynamics of Elastic Materials with Heat Conduction and Viscosity*, Arch. Ratl. Mech. Anal., 13, 167–178 (1963).

[11] Courant, R., and D. Hilbert, *Methods of Mathematical Physics*, Interscience, New York, 1952.

[12] Da Vinci, Leonardo, *Opere*, Ediz. Bibl. Vatican, 1916. The work on sound waves dates back to about 1500.

[13] Ding, J., and D. Gidaspow, *A Bubbling Fluidization Model using Kinetic Theory of Granular Flow*, AIChEAJ, 36, 523–538 (1990).

[14] Drew, D., *Mathematical Modeling of Two-Phase Flow*, Ann. Rev. Fluid Mech., 15, 261–291 (1983).

[15] Foscolo, P.U., and L.G. Gibilaro, *Fluid Dynamic Stability of Fluidized Suspensions: the Particle Bed Model*, Chem. Eng. Sci., 42, 1489–1500 (1987).

[16] Foscolo, P.U., L.G. Gibilaro and S. Rapagna', *Infinitesimal and Finite Voidage Perturbations in the Compressible Particle Phase Description of a Fluidized Bed*, AIChE Symp. Series, *Fluidization and Fluid-Particle Systems*, 1995.

[17] Goldreich, P., and S.A. Tremaine, *The Velocity Dispersion in Saturn's Rings*, Icarus, 34, 227–239 (1978).

[18] Haff, P.K., *Grain Flow as a Fluid Mechanical Phenomenon*, J. Fluid Mech., 134, 401–430 (1983).

[19] Hugoniot, H., *Mémoire sur la Propagation du Mouvement dans les Corps et spécialemnet dan les Gaz Parfaits*, J. Ecole Polyt. 58, 1–17 (1888).

[20] Hugoniot, H., *Mémoire sur la Propagation du Mouvement dans un Fluide Indefini*, J. Math. Pur. Appl., 3, 477–491; 4, 153–167 (1887).

[21] Jackson, R., *The Flow of Granular Materials and Aerated Granular Material*, J. Rheol., 30, 907–930 (1986).

[22] Jean, R.H. and L.S. Fan, *On the Model Equations of Gibilaro and Foscolo with Corrected Buoyancy Forces*, Powder Techn., 62, 201–205 (1992).

[23] Jenkins, J.T. and S.B. Savage, *A Theory for the Rapid Flow of Identical, Smooth, Nearly Elastic Spherical Particles*, J. Fluid Mech., 130, 187–202 (1983).

[24] Joseph, D.D., Lundgren, T.S., (with an Appendix by R. Jackson and D.A. Saville), *Ensemble Averaged and Mixture Theory Equations for Incompressible Fluid-Particle Suspensions*, Int. J. Multiphase Flow, 16, 35–42 (1990).

[25] Laplace, P.S., "Oeuvres", (Imprimerie Royale, Paris 1846). The work on the speed of sound dates back to 1816.

[26] Lax, P.D., *Hyperbolic Systems of Conservation Laws and the Mathematical Theory of Shock Waves*, SIAM, Philadelphia, 1973.

[27] Lundgren, T.S., *Slow Flow through Stationary Random Beds and Suspensions of Spheres*, J. Fluid Mech., 51, 273–299 (1972).

[28] Maddox, J., *Towards Unequal Partition of Energy?*, Nature, 374 (6517), 11, 1995.

[29] Maxwell, J.C., *On the Dynamical Theory of Gases*, Phil. Trans. Roy. Soc. London, 157, 49–98 (1867).

[30] Mc Tigue, D.F., and J.T. Jenkins, *Channel Flow of Concentrated Suspensions*, in: *Advances in Micromechanics of Granular Materials*, H.H. Shen, M. Statake, M. Mehrabadi, C.S. Chang, C.S. Campbell Eds., Elsevier, Amsterdam, 1992.

[31] Morse, P.M., and H. Feshbach, *Methods of Theoretical Physics*, McGraw Hill, New York, 1953.

[32] Mutsers, S.M.P., and K. Rietema, *The Effect of Interparticle Forces on the Expansion of a Homogeneous Gas-Fluidized Bed*, Powder Techn., 18, 239–248 (1977).

[33] Nayfeh, A.H., *Perturbation Methods*, J. Wiley, New York, 1973.

[34] Newton, I., *Philosophiae Naturalis Principia Mathematica*, Streeter, London 1687.

[35] Ocone, R., and G. Astarita, *A Pseudo-Thermodynamic Theory of Granular Flow Rheology*, J. Rheol., 37, 727–742 (1993).

[36] Ocone, R., and G. Astarita, *On Waves of Particulate Phase Pressure in Granular Materials*, J. Rheol., 38, 129–139 (1994).

[37] Ocone, R., and G. Astarita, *Compression and Rarefaction Waves in Granular Flow*, Powder Techn., 82, 231–237 (1995a).

[38] Ocone, R., and G. Astarita, *Energy Partitions*, Nature, 375 (6530), 254 (1995b).

[39] Ocone, R., and G. Astarita, *Grain Inertia and Fluid Viscosity Dominated Granular Flow Rheology: A Thermodynamics Analysis*, Rheol. Acta, 34, 323–328 (1995c).

[40] Ogawa, S., *Multitemperature Theory of Granular Materials*, in Proc. of the US-Japan Symp. on Cont. Mech. and Statist. Appr. in Mech. of Granular Materials, S.C. Cowin and M. Satake Eds., Gukujatsu Bunken Fukyukai, 1978.

[41] Rhee, K.H., R. Aris, and N.R. Amundson, *First Order Partial Differential Equations*, Prentice Hall, Englewood Cliffs, 1989.

[42] Richtmeyer, R.D., and K.W. Morton, *Difference Methods for Initial Value Problems*, Interscience, New York, 1967.

[43] Rosensweig, R.E., *Magnetic Stabilization of the State of Uniform Fluidization*, Ind. Eng. Chem. Fund., 18, 260–269 (1979).

[44] Savage, S.B., *Streaming Motions in a Bed of Vibrationally Fluidized Dry Granular Material*, J. Fluid Mech., 194, 457–478 (1988).

[45] Truesdell, C.A., *Rational Thermodynamics*, 2nd Ed., Springer-Verlag, Berlin, 1984.

[46] Waterston, J.J., *The Physics of Media that are Composed of Free and Perfectly Elastic Molecules in a State of Motion*, Phil. Trans. Roy. Soc. London, A183, 1–79 (1893). Published posthumously, 48 years after its submission in 1845.

# EFFECTIVE MEDIA THEORY USING NEAREST NEIGHBOR PAIR DISTRIBUTIONS

D.A. DREW* AND H. MANDYAM*

**1. Introduction.** The behavior of heterogeneous materials is interesting on several levels. First, such materials are common in nature and our technological society. Their behavior is difficult to predict, to understand, and to explain. On a theoretical level, the nature of these materials leads naturally to questions of repeatability and averaging. The microscale problem is often describable by classical dynamical equations (Navier-Stokes, Fourier heat conduction, etc.), but using this information to obtain constitutive equations is difficult, and has led to the development of two theoretical techniques (viz., renormalization and effective media theory).

In this paper, we examine some of the models for the mechanics and heat conduction of a suspension of spheres. This paper has several goals. In part, it represents an attempt to summarize some of the techniques and results for heterogeneous materials. This summary is incomplete. This paper also introduces a new method of derivation of constitutive equations, using a nearest neighbor pair correlation function. This method is applied to heat conduction. It has not been applied to effective viscosity or drag.

**2. Ensemble averaging.** Here, we define the ensemble average, and give some results pertaining to its application to multicomponent flows.

Suppose we have a medium made up of rigid spheres in a matrix. Suppose further that the spheres are approximately equally likely to be found at any point in the matrix. Clearly there are a large number of possible positions that the spheres could occupy. It is difficult is to establish the exact locations of each of the spheres. Furthermore, the exact locations of each of the spheres is not of particular interest in many applications. For purposes of the solution of engineering problems, different arrangements of the spheres are indistinguishable. The engineering problem is often specified by a set of partial differential equations for fields inside and outside of the spheres, along with jump conditions across the boundary of the spheres. In addition, appropriate initial and/or boundary conditions must be specified for the problem. The solution of the governing equations on an arrangement of spheres is a realization of *the same system.* In this case and for this purpose, it is not possible to keep an exact accounting of the exact state of the system, *nor is it desirable.*

Indeed, it is a necessity for most useful prediction that the features that are of interest for a particular realization must be insensitive to small

* Department of Mathematical Science, Rensselaer Polytechnic Institute, Troy, NY 12180-3590.

changes in the the physical conditions that specify the problem. Were it otherwise, the dynamics of the system might not be repeatable, so that other researchers cannot verify the results.

We assume a set of possible realizations, called an *ensemble*. A reasonable ensemble for a flow problem might be the motions resulting from a set of initial conditions for the positions of a number of spheres in the flow. For a problem of conductivity, the ensemble consists of the temperature field in steady state conduction through the matrix and the spheres. We must assign probabilities to the realizations. Formally, then, *an ensemble is a set of solutions "possible" in the system.* Suppose $f$ is a field. We denote the realization by $\mu$, and denote the dependence of the field on the realization by $f(\mathbf{x}, t; \mu)$. We refer to the set $\mathcal{E}$, of all realizations $\mu$, as the *ensemble*. The average of $f$ is

$$\overline{f}(\mathbf{x}, t) = \int_{\mathcal{E}} f(\mathbf{x}, t; \mu)\, dm(\mu) \tag{2.1}$$

where $dm(\mu)$ is the probability (measure) of the realization $\mu$.

**2.1. Definition of $X_k$.** It is often desirable to isolate each component theoretically, even in the microscale description. To do this, we introduce the *component indicator function*, or *characteristic function*, $X_k(\mathbf{x}, t)$, defined by

$$X_k = \begin{cases} 1, & \text{if } \mathbf{x} \in k\,, \\ 0, & \text{otherwise.} \end{cases} \tag{2.2}$$

This function "picks out" component $k$, and ignores all other components and interfaces. This function is crucial to the description of multicomponent materials.

**3. Averaged equations.** We now present the averaged equations. The averaged equations governing each component can be obtained by multiplying the exact equations of conservation of mass, momentum, and energy by $X_k$ and averaging [Drew and Wallis, 1996]. The average equations are

**Mass**

$$\frac{\partial \alpha_k \rho_k}{\partial t} + \nabla \cdot \alpha_k \rho_k \mathbf{v}_k = 0 \tag{3.1}$$

**Momentum**

$$\frac{\partial \alpha_k \rho_k \mathbf{v}_k}{\partial t} + \nabla \cdot \alpha_k \rho_k \mathbf{v}_k \mathbf{v}_k = \nabla \cdot \alpha_k \left(\mathbf{T}_k + \mathbf{T}_k^{Re}\right) + \alpha_k \rho_k \mathbf{g} + \mathbf{M}_k \tag{3.2}$$

**Internal Energy**

$$\frac{\partial \alpha_k \rho_k u_k}{\partial t} + \nabla \cdot \alpha_k \rho_k \mathbf{v}_k u_k = \alpha_k \mathbf{T}_k : \nabla \mathbf{v}_k - \nabla \cdot \alpha_k (\mathbf{q}_k + \mathbf{q}_k^{Re}) + \alpha_k \rho_k r_k + E_k + \alpha_k D_k - \alpha_k P_k . \tag{3.3}$$

The average of the indicator function is

$$\alpha_k = \overline{X_k} . \tag{3.4}$$

The remaining variables are defined in terms of weighted, or conditional, averages. We show here those for which we provide constitutive equations using averaging techniques. A complete list is available in Drew and Wallis [1996].

The averaged stress in component $k$ is

$$\mathbf{T}_k = \overline{X_k \mathbf{T}} / \alpha_k , \tag{3.5}$$

and the averaged energy flux in component $k$ is

$$\mathbf{q}_k = \overline{X_k \mathbf{q}} / \alpha_k . \tag{3.6}$$

The interfacial momentum source to component $k$ is defined by

$$\mathbf{M}_k = -\overline{\mathbf{T} \cdot \nabla X_k} , \tag{3.7}$$

the interfacial heat source to component $k$ is defined by

$$E_k = \overline{\mathbf{q} \cdot \nabla X_k} , \tag{3.8}$$

the interfacial work in component $k$ is defined by

$$W_k = -\overline{\mathbf{T} \cdot \mathbf{v} \cdot \nabla X_k} . \tag{3.9}$$

The Reynolds stress in component $k$ is

$$\mathbf{T}_k^{Re} = -\overline{X_k \rho \mathbf{v}_k' \mathbf{v}_k'} / \alpha_k . \tag{3.10}$$

and the fluctuation (Reynolds) internal energy flux is

$$\mathbf{q}_k^{Re} = \overline{X_k \rho \mathbf{v}_k' u_k'} / \alpha_k . \tag{3.11}$$

The dissipation is

$$D_k = \frac{\overline{X_k \mathbf{T} : \nabla \mathbf{v}_k'}}{\alpha_k} \tag{3.12}$$

**3.1. Jump conditions.** We shall assume that the interface between the fluid and the particles is "inert", in that it exerts no surface tension, nor contributes any energy. Then balance of momentum and energy across the interface between the components yields the following conditions:

**Momentum**

$$\mathbf{M}_1 + \mathbf{M}_2 = 0\,, \tag{3.13}$$

**Energy**

$$E_1 + W_1 + E_2 + W_2 = 0\,. \tag{3.14}$$

**3.2. Closure.** The average field equations and jump conditions must be supplemented by constitutive equations which in some sense replace the information lost in the averaging process. There are several ways to proceed; in this paper, we shall use averaging techniques to derive constitutive equations for effective heat conduction and viscosity.

**4. Statistical mechanics.** Consider a system consisting of $N$ discrete units, which may be droplets, particles, or bubbles. For simplicity, we assume that the particles are identical spheres and are therefore indistinguishable. Furthermore, we need not include information about the particle sizes or orientations in this description. In order to describe the randomness in the system, assume that there is a "master" distribution function

$$f^{(N)}(\mathbf{z}_1, \mathbf{v}_1, \mathbf{z}_2, \mathbf{v}_2, \ldots, \mathbf{z}_N, \mathbf{v}_N, t)\,. \tag{4.1}$$

Roughly speaking, the distribution function $f^{(N)}$ has the interpretation that

$$f^{(N)}\, d\mathbf{z}_1\, d\mathbf{v}_1 \;\cdots d\mathbf{z}_N\, d\mathbf{v}_N \tag{4.2}$$

is the probability of finding a particle within $d\mathbf{z}_1$ of $\mathbf{z}_1$ with velocity within $d\mathbf{v}_1$ of $\mathbf{v}_1$ and finding a particle within $d\mathbf{z}_2$ of $\mathbf{z}_2$ with velocity within $d\mathbf{v}_2$ of $\mathbf{v}_2$, *etc.*, at time $t$, during a process $\mu$. Note that this assumes that each particle is identified with a number, and that we can distinguish particle $i$ from particle $j$, for $i \neq j$.

The other particle distribution functions can be defined similarly. For example, the single particle distribution function $f^{(1)}(\mathbf{z}, \mathbf{v}, t)$ is defined by taking $f^{(1)}\, d\mathbf{z}\, d\mathbf{v}$ as the probability of finding a particle within $d\mathbf{z}$ of $\mathbf{z}$, with velocity within $d\mathbf{v}$ of $\mathbf{v}$. The interpretation in terms of sub-ensembles follows. It is interesting to note that we have

$$f^{(1)} = \int\!\!\int \cdots \int f^{(N)}\, d\mathbf{z}_2 d\mathbf{v}_2 \cdots d\mathbf{z}_N d\mathbf{v}_N\,. \tag{4.3}$$

*Particle Distribution Function.* For the purposes of this paper, we shall suppress the dependence of $f^{(N)}$ on the particle velocities. Then

$$f^{(N)}(\mathbf{z}_1, \mathbf{z}_2, \ldots, \mathbf{z}_N)$$

is the distribution function of the positions of the centers. By symmetry, if

$$\{i_1, i_2, \ldots, i_N\},$$

is any rearrangement of the particles, then

$$f^{(N)}(\mathbf{z}_1, \mathbf{z}_2, \ldots, \mathbf{z}_N) = f^{(N)}(\mathbf{z}_{i_1}, \mathbf{z}_{i_2}, \ldots, \mathbf{z}_{i_N}).$$

**4.1. Nearest neighbor pair distribution function.** For each point $\mathbf{x}$, there is a rearrangement of the numbering of the spheres in order of closeness to the point $\mathbf{x}$. That is, sphere number $i_1$, at location $\mathbf{z}'_1 = \mathbf{z}_{i_1}$, is closest to $\mathbf{x}$, sphere number $i_2$, at location $\mathbf{z}'_2 = \mathbf{z}_{i_2}$, is next closest, and so on. Note that the order changes as the point $\mathbf{x}$ changes, and will change as $t$ changes. The distribution function in terms of the spheres numbered in order of closeness to the point $\mathbf{x}$, at time $t$, can be found in terms of the original distribution function by

$$\hat{f}^{(N)}(\mathbf{x}, t; \mathbf{z}'_1, \ldots, \mathbf{z}'_N) = f^{(N)}(\mathbf{z}_1, \ldots, \mathbf{z}_N). \tag{4.4}$$

For simplicity, we shall suppress the notation for the dependence on $t$. It should be realized, however, that the closest particle distribution can change in space and time.

Note that ties in order of closeness can occur. Situations involving ties are events of low probability, and in most situations the way in which the tie is broken will not affect the results.

Given a distribution of particles at $\mathbf{z}_1, \mathbf{z}_2, \ldots, \mathbf{z}_N$, suppose $\phi$ is some field that depends on position and time, and also the arrangement of the spheres:

$$\phi(\mathbf{x}|\mathbf{z}_1, \mathbf{z}_2, \ldots, \mathbf{z}_N)$$

We wish to perform the average by multiplying $\phi$ by $f^{(N)}$ and integrating over all possible particle positions. Thus,

$$\overline{\phi} = \int_{\mathbf{z}_1} \int_{\mathbf{z}_2} \cdots \int_{\mathbf{z}_N} f^{(N)}(\mathbf{z}_1, \ldots, \mathbf{z}_N)\, \phi(\mathbf{x}|\mathbf{z}_1, \ldots, \mathbf{z}_N)\, d\mathbf{z}_1 \ldots d\mathbf{z}_N. \tag{4.5}$$

It is also useful to define conditional averages. Batchelor [1] and Hinch [2] write (4.5) as

$$\begin{aligned} \overline{\phi} &= \int_{\mathbf{z}_1} f^{(1)}(\mathbf{z}_1)\overline{\phi}^{(1)}(\mathbf{x}|\mathbf{z}_1)\, d\mathbf{z}_1, \\ &= \int_{\mathbf{z}_1} \int_{\mathbf{z}_2} f^{(2)}(\mathbf{z}_1, \mathbf{z}_2)\, \overline{\phi}^{(2)}(\mathbf{x}|\mathbf{z}_1, \mathbf{z}_2)\, d\mathbf{z}_1 d\mathbf{z}_2, \end{aligned} \tag{4.6}$$

*etc.*, where

$$\overline{\phi}^{(1)}(\mathbf{x}|\mathbf{z}_1) = \frac{1}{f^{(1)}(\mathbf{z}_1)} \int_{\mathbf{z}_2} \int_{\mathbf{z}_3} \cdots \int_{\mathbf{z}_N} f^{(N)}(\mathbf{z}_1, \mathbf{z}_2, \ldots, \mathbf{z}_N) \times \phi(\mathbf{x}|\mathbf{z}_1, \mathbf{z}_2, \ldots, \mathbf{z}_N)\, d\mathbf{z}_2\, d\mathbf{z}_3 \ldots d\mathbf{z}_N \tag{4.7}$$

is the conditionally averaged field, averaged on the condition that there is a sphere with its center at $\mathbf{z}_1$. Here, it is assumed that the appropriate definition for $f^{(1)}(\mathbf{z}_1)$ is the unconditional density function for any sphere center being at $\mathbf{z}_1$,

$$f^{(1)}(\mathbf{z}_1) = \int_{\mathbf{z}_2} \int_{\mathbf{z}_3} \cdots \int_{\mathbf{z}_N} f^{(N)}(\mathbf{z}_1, \mathbf{z}_2, \ldots, \mathbf{z}_N)\, d\mathbf{z}_2 d\mathbf{z}_3 \ldots d\mathbf{z}_N . \tag{4.8}$$

Similarly,

$$\overline{\phi}^{(2)}(\mathbf{x}|\mathbf{z}_1, \mathbf{z}_2) = \frac{1}{f^{(2)}(\mathbf{z}_1, \mathbf{z}_2)} \int_{\mathbf{z}_3} \cdots \int_{\mathbf{z}_N} f^{(N)}(\mathbf{z}_1, \mathbf{z}_2, \ldots, \mathbf{z}_N) \times \phi(\mathbf{x}|\mathbf{z}_1, \mathbf{z}_2, \ldots, \mathbf{z}_N)\, d\mathbf{z}_3 \ldots d\mathbf{z}_N , \tag{4.9}$$

and

$$f^{(2)}(\mathbf{z}_1, \mathbf{z}_2) = \int_{\mathbf{z}_3} \int_{\mathbf{z}_4} \cdots \int_{\mathbf{z}_N} f^{(N)}(\mathbf{z}_1, \mathbf{z}_2, \ldots, \mathbf{z}_N)\, d\mathbf{z}_3 d\mathbf{z}_4 \ldots d\mathbf{z}_N , \tag{4.10}$$

*etc.* The distributions $f^{(i)}$ are called the $i$-particle distribution function.

In terms of the distribution of closest spheres, the average velocity is

$$\overline{\phi} = \int_{\mathbf{z}'_1} \int_{\mathbf{z}'_2} \cdots \int_{\mathbf{z}'_N} \hat{f}^{(N)}(\mathbf{x}, \mathbf{z}'_1, \mathbf{z}'_2, \ldots, \mathbf{z}'_N) \phi(\mathbf{x}|\mathbf{z}'_1, \mathbf{z}'_2, \ldots, \mathbf{z}'_N)\, d\mathbf{z}'_1 d\mathbf{z}'_2 \ldots d\mathbf{z}'_N . \tag{4.11}$$

There are corresponding conditional averages and distributions. Consider

$$\begin{aligned} \overline{\phi}(\mathbf{x}) &= \int_{\mathbf{z}'_1} \hat{f}^{(1)}(\mathbf{x}, \mathbf{z}'_1)\, \hat{\overline{\phi}}^{(1)}(\mathbf{x}|\mathbf{z}'_1)\, d\mathbf{z}'_1 , \\ &= \int_{\mathbf{z}'_1} \int_{\mathbf{z}'_2} \hat{f}^{(2)}(\mathbf{x}, \mathbf{z}'_1, \mathbf{z}'_2)\, \hat{\overline{\phi}}^{(2)}(\mathbf{x}|\mathbf{z}'_1, \mathbf{z}'_2)\, d\mathbf{z}'_1 d\mathbf{z}'_2 , \end{aligned} \tag{4.12}$$

*etc.*, where

$$\hat{\overline{\phi}}^{(1)}(\mathbf{x}|\mathbf{z}'_1) = \frac{1}{\hat{f}^{(1)}(\mathbf{x}, \mathbf{z}'_1)} \int_{\mathbf{z}'_2} \int_{\mathbf{z}'_3} \cdots \int_{\mathbf{z}'_N} \hat{f}^{(N)}(\mathbf{x}, \mathbf{z}'_1, \mathbf{z}'_2, \ldots, \mathbf{z}'_N) \times \phi(\mathbf{x}|\mathbf{z}'_1, \mathbf{z}'_2, \ldots, \mathbf{z}'_N)\, d\mathbf{z}'_2 d\mathbf{z}'_3 \ldots d\mathbf{z}'_N \tag{4.13}$$

is the conditionally averaged velocity, averaged on the condition that there is a sphere with its center at $\mathbf{z}'_1$.

*Computing the Distribution Functions.* We develop a technique for computing the functions $\hat{f}^{(1)}$ and $\hat{f}^{(2)}$. A *uniform particle concentration* is assumed so that $f^{(1)}(\mathbf{z}_1)$ is constant. If the volumetric concentration is given as $\alpha_p(\mathbf{x},t)$, then

$$f^{(1)}(\mathbf{z}_1) = \frac{\alpha_p(\mathbf{x},t)}{\frac{4}{3}\pi a^3}.$$

For the distributions of closest particles, note that $\hat{f}^{(1)}(\mathbf{x},\mathbf{z}_1')\,d\mathbf{z}_1'$ is the probability that the sphere having its center within $d\mathbf{z}_1'$ of $\mathbf{z}_1'$ is the closest to the point $\mathbf{x}$, and

$$\hat{f}^{(2)}(\mathbf{x},\mathbf{z}_1',\mathbf{z}_2')\,d\mathbf{z}_1'\,d\mathbf{z}_2' \tag{4.14}$$

is the probability that the sphere having its center within $d\mathbf{z}_1'$ of $\mathbf{z}_1'$ is the closest to the point $\mathbf{x}$ *and* the sphere having its center within $d\mathbf{z}_2'$ of $\mathbf{z}_2'$ is the next closest to the point $\mathbf{x}$, and so on. These distributions are the one-particle and two-particle nearest neighbor distribution function for spheres closest to the space point $\mathbf{x}$.

In order to compute an expression for $\hat{f}^{(1)}(\mathbf{x},\mathbf{z}_1')$, consider the probability that there is no sphere center within a distance $r$ of $\mathbf{x}$ and that there is a sphere center within the shell between $r$ and $r+dr$. Then it is clear that

$$\hat{f}^{(1)}(\mathbf{x},\mathbf{z}_1')\,d\mathbf{z}_1' \tag{4.15}$$

is the probability that there is no sphere within $r$ of $\mathbf{x}$ *and* there is one sphere within $d\mathbf{z}_1'$ of $\mathbf{z}_1'$. Now, the probability that there is no sphere within $r$ of $\mathbf{x}$ and the probability that the nearest sphere is within $r$ of $\mathbf{x}$ sum to unity. Thus,

$$\hat{f}^{(1)}(\mathbf{x},\mathbf{z}_1')\,d\mathbf{z}_1' = \left(1 - \int\limits_{|\mathbf{x}-\mathbf{z}_1''|\le|\mathbf{x}-\mathbf{z}_1'|} \hat{f}^{(1)}(\mathbf{x},\mathbf{z}_1'')\,d\mathbf{z}_1''\right) f^{(1)}(\mathbf{z}_1')\,d\mathbf{z}_1'. \tag{4.16}$$

If we take $f^{(1)}(\mathbf{z}_1')$ =constant, we see that $\hat{f}^{(1)}$ is a function only of $r_1 = |\mathbf{x}-\mathbf{z}_1'|$. Then, noting that if we replace $d\mathbf{z}_1'$ by $4\pi(r_1')^2 dr_1'$, the left side of (4.16) is the differential of the integral on the right. Then we have

$$\frac{d}{dr_1}\ln\left(1 - 4\pi\int_0^{r_1} \hat{f}^{(1)}(r_1')(r_1')^2\,dr_1'\right) = -4\pi f^{(1)} r_1^2.$$

Integrating,

$$\hat{f}^{(1)}(\mathbf{x},\mathbf{z}_1') = C_1 \exp\left(-\frac{4}{3}\pi f^{(1)}|\mathbf{x}-\mathbf{z}_1'|^3\right). \tag{4.17}$$

where $C_1$ is a constant. We evaluate the constant by requiring

$$\int \hat{f}^{(1)}\, d\mathbf{z}_1 = 1 \tag{4.18}$$

Thus,

$$\hat{f}^{(1)}(\mathbf{x}, \mathbf{z}_1') = f^{(1)} \exp\left(-\frac{4}{3}\pi f^{(1)} r_1^3\right) . \tag{4.19}$$

A similar derivation can be carried out for the two-particle distribution. This distribution is useful for proving convergence results for averages of one-particle distributions.

The distribution function for two closest particles, $\hat{f}^{(2)}(\mathbf{x}, \mathbf{z}_1', \mathbf{z}_2')$ vanishes for $|\mathbf{x} - \mathbf{z}_2'| \leq |\mathbf{x} - \mathbf{z}_1'|$. For $|\mathbf{x} - \mathbf{z}_2'| > |\mathbf{x} - \mathbf{z}_1'|$, $\hat{f}^{(2)}(\mathbf{x}, \mathbf{z}_1', \mathbf{z}_2')\, d\mathbf{z}_1'\; d\mathbf{z}_2'$ is the probability that the sphere having its center within $d\mathbf{z}_1'$ of $\mathbf{z}_1'$ is the closest to the point $\mathbf{x}$ *and* the sphere within $d\mathbf{z}_2'$ of $\mathbf{z}_2'$ is the next closest to the point $\mathbf{x}$. This can be calculated in terms of the conditional probability that the next closest sphere to $\mathbf{x}$ is within $d\mathbf{z}_2'$ of $\mathbf{z}_2'$, given that there is a sphere at $\mathbf{z}_1'$. This, in turn, can be calculated in a manner similar to the closest sphere density above. It is clear that the conditional probability,

$$\hat{f}^{(2)}(\mathbf{x}, \mathbf{z}_2' | \mathbf{z}_1')\, d\mathbf{z}_2' ,$$

satisfies

$$\hat{f}^{(2)}(\mathbf{x}, \mathbf{z}_2' | \mathbf{z}_1')\, d\mathbf{z}_2' = 0 , \tag{4.20}$$

for $|\mathbf{x} - \mathbf{z}_2'| < |\mathbf{x} - \mathbf{z}_1'|$. For $|\mathbf{x} - \mathbf{z}_2'| > |\mathbf{x} - \mathbf{z}_1'|$ we have

$$\hat{f}^{(2)}(\mathbf{x}, \mathbf{z}_2' | \mathbf{z}_1')\, d\mathbf{z}_2' = \left(1 - \int\limits_{|\mathbf{x}-\mathbf{z}_1'| \leq |\mathbf{x}-\mathbf{z}_2''| \leq |\mathbf{x}-\mathbf{z}_2'|} \hat{f}^{(2)}(\mathbf{x}, \mathbf{z}_1'', \mathbf{z}_2'')\, d\mathbf{z}_2''\right) f^{(1)}\, d\mathbf{z}_2' . \tag{4.21}$$

Then we see that $\hat{f}^{(2)}$ is a function only of $r_2 = |\mathbf{x} - \mathbf{z}_2'|$. Thus,

$$\hat{f}^{(2)}(r_2 | \mathbf{z}_1') = \left(1 - 4\pi \int_{r_1}^{r_2} \hat{f}^{(2)}(r_2' | \mathbf{z}_1')(r_2')^2\, dr_2'\right) f^{(1)} . \tag{4.22}$$

Solving for $\hat{f}^{(2)}$ yields

$$\hat{f}^{(2)}(\mathbf{x}, \mathbf{z}_2' | \mathbf{z}_1') = C_2 \exp\left[-\frac{4\pi}{3} f^{(1)} r_2^3\right] . \tag{4.23}$$

where $C_2$ is a constant of integration. It is difficult to evaluate the constant of integration, because it depends on the assumptions made about exclusion or overlapping of spheres. If we assume that the spheres can overlap ("soft sphere" model), then

$$C_2 = f^{(1)} \exp\left[\frac{4\pi}{3} f^{(1)} (|\mathbf{x} - \mathbf{z}_1'|)^3\right] . \tag{4.24}$$

Then the conditional probability density is given by

$$\hat{f}^{(2)}(\mathbf{x}, \mathbf{z}_2'|\mathbf{z}_1') = f^{(1)} \exp\left\{\alpha_p \left[\left(\frac{|\mathbf{x} - \mathbf{z}_1'|}{a}\right)^3 - \left(\frac{|\mathbf{x} - \mathbf{z}_2'|}{a}\right)^3\right]\right\} . \tag{4.25}$$

If, instead, we consider the "hard sphere" model, the calculation of the constant is difficult. However, if we assume that the nearest sphere is centered on the space point $\mathbf{x}$, then we have

$$C_2 = f^{(1)} \exp\left[8\alpha_p + \frac{4\pi}{3} f^{(1)} (|\mathbf{x} - \mathbf{z}_1'|)^3\right] . \tag{4.26}$$

Then the conditional probability density is given by

$$\hat{f}^{(2)}(\mathbf{x}, \mathbf{z}_2'|\mathbf{z}_1') = f^{(1)} \exp\left\{\alpha_p \left[8 + \left(\frac{|\mathbf{x} - \mathbf{z}_1'|}{a}\right)^3 - \left(\frac{|\mathbf{x} - \mathbf{z}_2'|}{a}\right)^3\right]\right\} . \tag{4.27}$$

*Convergence Results.* The averages discussed above are used to derive results about constitutive equations from information about exact flows. There are difficulties regarding this procedure, however.

To see this, consider a function

$$g_1(\mathbf{x}, t|\mathbf{z}_1', \ldots, \mathbf{z}_N') = g(\mathbf{x}, t|\mathbf{z}_1') + O\left(\frac{1}{|\mathbf{x} - \mathbf{z}_2'|^\nu}\right) ,$$

where $g_1$ is of order 1. We note the ordinary (*i.e.*, *unordered*) expression for the probability density leads to averages that have convergence problems in unbounded flows. The average of $g$ is given by

$$\overline{g}(\mathbf{x}, t) = \int\limits_{\mathbf{z}_1} \ldots \int\limits_{\mathbf{z}_N} f^{(N)}(\mathbf{x}, \mathbf{z}_1, \ldots, \mathbf{z}_N)\; g(\mathbf{x}, t|\mathbf{z}_1, \ldots, \mathbf{z}_N)\, d\mathbf{z}_1 \ldots d\mathbf{z}_N ,$$

$$= \int\limits_{\mathbf{z}_1} f^{(1)}(\mathbf{z}_1)\; g_1(\mathbf{x}, t|\mathbf{z}_1)\, d\mathbf{z}_1 + E , \tag{4.28}$$

where the error $E$ is due to all the spheres except the one at $\mathbf{z}_1$. If $f^{(1)}$ is constant, the integral will not converge, nor will it converge if the distribution does not decrease sufficiently rapidly.

In terms of the "ordered" distributions, the average of $g$ is given by

$$\hat{\overline{g}}(\mathbf{x}, t) = \int\limits_{\mathbf{z}_1'} \ldots \int\limits_{\mathbf{z}_N'} \hat{f}^{(N)}(\mathbf{x}, \mathbf{z}_1', \ldots, \mathbf{z}_N')\, g(\mathbf{x}, t|\mathbf{z}_1', \ldots, \mathbf{z}_N')\, d\mathbf{z}_1' \ldots d\mathbf{z}_N' ,$$

$$= \int\limits_{\mathbf{z}_1'} \hat{f}^{(1)}(\mathbf{x}, \mathbf{z}_1')\; g_1(\mathbf{x}, t|\mathbf{z}_1')\, d\mathbf{z}_1' + \hat{E} , \tag{4.29}$$

where $\hat{E}$ is an error, which is of order

$$\int_{\mathbf{z}_1'}\int_{\mathbf{z}_2'} \hat{f}^{(2)}(\mathbf{x},\mathbf{z}_1',\mathbf{z}_2')\left(\frac{1}{|\mathbf{x}-\mathbf{z}_2'|^{\nu}}\right) d\mathbf{z}_1'\, d\mathbf{z}_2' .$$

Substituting $\hat{f}^{(2)}$ from Equation (4.27) shows that the error is of order

$$\int_0^{\infty}\int_{r_1}^{\infty} (f^{(1)})^2 \exp(-\frac{4}{3}\pi f^{(1)} r_2^3) 4\pi r_1^3\, 4\pi r_2^{3-\nu}\, dr_1 dr_2 . \tag{4.30}$$

If we make the change of variables $r_1' = (f^{(1)})^{1/3} r_1$ and $r_2' = (f^{(1)})^{1/3} r_2$, it is apparent that $\hat{E} = O\left((f^{(1)})^{\nu/3}\right)$. Thus, the one-sphere calculation can give results that are accurate for dilute flows, where $f^{(1)}$ is small.

**5. Constitutive equations for heat conduction and Stokes flow.** In this Section, we shall present a number of constitutive equations for flow and heat conduction to illustrate the methods of computation and the final equations.

We shall use two approximations to obtain equations for the heat conduction and viscous behavior of a suspension of spheres. First, we shall ignore all effects of mechanical energy on the thermal processes. Second, we shall assume that inertial forces in the mechanical equations are small.

**5.1. Averaged equations.**

*Heat Equation.* First, for the heat equation, we have

$$\frac{\partial \alpha_k \rho_k u_k}{\partial t} + \nabla \cdot \alpha_k \rho_k \mathbf{v}_k u_k = -\nabla \cdot \alpha_k \mathbf{q}_k + \alpha_k \rho_k r_k + E_k , \tag{5.1}$$

where

$$\alpha_k \mathbf{q}_k = \overline{X_k \mathbf{q}} = \int \dots \int f^{(N)}(\mathbf{z}_1, \dots, \mathbf{z}_N) \mathbf{q}(\mathbf{x}, t|\mathbf{z}_1, \dots \mathbf{z}_N) d\mathbf{z}_1 \dots d\mathbf{z}_N \tag{5.2}$$

and

$$\begin{aligned} E_k &= -\overline{\mathbf{q} \cdot \nabla X_k} \\ &= \int \dots \int f^{(N)}(\mathbf{z}_1, \dots, \mathbf{z}_N) \mathbf{n} \cdot \mathbf{q}(\mathbf{x}, t|\mathbf{z}_1, \dots \mathbf{z}_N) \delta(\mathbf{x} - \mathbf{x}_i) d\mathbf{z}_1 \dots d\mathbf{z}_N \end{aligned} \tag{5.3}$$

where $\delta(\mathbf{x} - \mathbf{x}_i)$ is the Dirac delta function for the interface.

*Momentum Equation.* The averaged mechanical equations for this system can be obtained by ignoring the inertia terms in the averaged balance equations. The equation of conservation of mass for component $k$ is

$$\frac{\partial \alpha_k}{\partial t} + \nabla \cdot (\alpha_k \mathbf{v}_k) = 0 . \tag{5.4}$$

The equation of balance of momentum for component $k$ is

$$\nabla \cdot \alpha_k \mathbf{T}_k + \mathbf{M}_k + \alpha_k \rho_k \mathbf{g}_k = 0\,, \tag{5.5}$$

where

$$\alpha_k \mathbf{T}_k = \overline{X_k \mathbf{T}} = \int \dots \int f^{(N)}(\mathbf{z}_1, \dots, \mathbf{z}_N) \mathbf{T}(\mathbf{x}, t | \mathbf{z}_1, \dots \mathbf{z}_N) d\mathbf{z}_1 \dots d\mathbf{z}_N \tag{5.6}$$

and

$$\begin{aligned} \mathbf{M}_k &= -\overline{\mathbf{T} \cdot \nabla X_k} \\ &= \int \dots \int f^{(N)}(\mathbf{z}_1, \dots, \mathbf{z}_N) \mathbf{n} \cdot \mathbf{T}(\mathbf{x}, t | \mathbf{z}_1, \dots \mathbf{z}_N) \delta(\mathbf{x} - \mathbf{x}_i) d\mathbf{z}_1 \dots d\mathbf{z}_N\,. \end{aligned} \tag{5.7}$$

Note that $\mathbf{M}_p + \mathbf{M}_f = 0$.

Adding (5.4) for $k = p$ and $k = f$ gives

$$\nabla \cdot (\alpha_f \mathbf{v}_f + \alpha_p \mathbf{v}_p) = 0\,. \tag{5.8}$$

Adding the momentum equations (5.5) for $k = p$ and $k = f$ gives

$$\nabla \cdot (\alpha_f \mathbf{T}_f + \alpha_p \mathbf{T}_p) + (\alpha_f \rho_f \mathbf{g}_f + \alpha_p \rho_p \mathbf{g}_p) = 0\,. \tag{5.9}$$

### 5.1.1. Conditionally averaged equations.

*Heat Equation.* If we average over all the positions of the spheres *except* the one at $\mathbf{z}$, we obtain the conditional averaged heat equations

$$\begin{aligned} &\frac{\partial \alpha_k^{(1)} \rho_k u_k^{(1)}}{\partial t} + \nabla \cdot \alpha_k^{(1)} \rho_k \mathbf{v}_k^{(1)} u_k^{(1)} \\ &\qquad = -\nabla \cdot \alpha_k^{(1)} (\mathbf{q}_k^{(1)}) + \alpha_k^{(1)} \rho_k r_k^{(1)} + E_k^{(1)}\,, \end{aligned} \tag{5.10}$$

where

$$\alpha_k^{(1)} \mathbf{q}_k^{(1)} = \int \dots \int f^{(N)}(\mathbf{z}_1, \dots, \mathbf{z}_N) \mathbf{q}(\mathbf{x}, t | \mathbf{z}_1, \dots \mathbf{z}_N) d\mathbf{z}_2 \dots d\mathbf{z}_N \tag{5.11}$$

and

$$E_k^{(1)} = \int \dots \int f^{(N)}(\mathbf{z}_1, \dots, \mathbf{z}_N) \mathbf{n} \cdot \mathbf{q}(\mathbf{x}, t | \mathbf{z}_1, \dots \mathbf{z}_N) \delta(\mathbf{x} - \mathbf{x}_i) d\mathbf{z}_2 \dots d\mathbf{z}_N \tag{5.12}$$

Similarly, the conditionally averaged equation keeping two spheres fixed is

$$\begin{aligned} &\frac{\partial \alpha_k^{(2)} \rho_k u_k^{(2)}}{\partial t} + \nabla \cdot \alpha_k^{(2)} \rho_k \mathbf{v}_k^{(2)} u_k^{(2)} \\ &\qquad = -\nabla \cdot \alpha_k^{(2)} (\mathbf{q}_k^{(2)}) + \alpha_k^{(2)} \rho_k r_k^{(2)} + E_k^{(2)}\,, \end{aligned} \tag{5.13}$$

where

$$\alpha_k^{(2)}\mathbf{q}_k^{(2)} = \int \cdots \int f^{(N)}(\mathbf{z}_1, \ldots, \mathbf{z}_N)\mathbf{q}(\mathbf{x}, t|\mathbf{z}_1, \ldots \mathbf{z}_N)d\mathbf{z}_3 \ldots d\mathbf{z}_N \tag{5.14}$$

and

$$E_k^{(2)} = \int \cdots \int f^{(N)}(\mathbf{z}_1, \ldots, \mathbf{z}_N)\mathbf{n}\cdot\mathbf{q}(\mathbf{x}, t|\mathbf{z}_1, \ldots \mathbf{z}_N)\delta(\mathbf{x}-\mathbf{x}_i)d\mathbf{z}_3 \ldots d\mathbf{z}_N \tag{5.15}$$

*Momentum Equation.* The conditionally averaged momentum equations can be written as

$$\nabla \cdot \alpha_k^{(1)}\mathbf{T}_k^{(1)} + \mathbf{M}_k^{(1)} + \alpha_k^{(1)}\rho_k\mathbf{g}_k = 0\,. \tag{5.16}$$

where

$$\alpha_k^{(1)}\mathbf{T}_k^{(1)} = \int \cdots \int f^{(N)}(\mathbf{z}_1, \ldots, \mathbf{z}_N)\mathbf{T}(\mathbf{x}, t|\mathbf{z}_1, \ldots \mathbf{z}_N)d\mathbf{z}_2 \ldots d\mathbf{z}_N \tag{5.17}$$

and

$$\mathbf{M}_k^{(1)} = -\int \cdots \int f^{(N)}(\mathbf{z}_1, \ldots, \mathbf{z}_N)\mathbf{n}\cdot\mathbf{T}(\mathbf{x}, t|\mathbf{z}_1, \ldots \mathbf{z}_N)\delta(\mathbf{x}-\mathbf{x}_i)d\mathbf{z}_2 \ldots d\mathbf{z}_N \tag{5.18}$$

$$\nabla \cdot \alpha_k^{(2)}\mathbf{T}^{(2)} + \mathbf{M}_k^{(2)} + \alpha_k^{(2)}\rho_k\mathbf{g}_k = 0\,, \tag{5.19}$$

where

$$\alpha_k^{(2)}\mathbf{T}_k^{(2)} = \int \cdots \int f^{(N)}(\mathbf{z}_1, \ldots, \mathbf{z}_N)\mathbf{T}(\mathbf{x}, t|\mathbf{z}_1, \ldots \mathbf{z}_N)d\mathbf{z}_3 \ldots d\mathbf{z}_N \tag{5.20}$$

and

$$\mathbf{M}_k^{(2)} = \int \cdots \int f^{(N)}(\mathbf{z}_1, \ldots, \mathbf{z}_N)\mathbf{n}\cdot\mathbf{T}(\mathbf{x}, t|\mathbf{z}_1, \ldots \mathbf{z}_N)\delta(\mathbf{x}-\mathbf{x}_i)d\mathbf{z}_3 \ldots d\mathbf{z}_N \tag{5.21}$$

**5.2. Heat conduction.** The calculation of the effective conductivity of a dispersion of spheres is a classical problem. Maxwell [3] was the first to calculate the effective conductivity of a random array of spheres. Maxwell computed the $O(\alpha_p)$ correction to the effective conductivity. One hundred years later, Jeffrey [4] derived the conductivity of a suspension of spheres through $O(\alpha_p^2)$.

Consider the heat flux around and through spheres of conductivity $\lambda_p$ in an infinite matrix of conductivity $\lambda_f$. The average temperature gradient is given by

$$\mathbf{G} = \nabla\overline{T} = \overline{(X_p + X_f)\nabla T} \tag{5.22}$$

Then the average heat flux is

$$\begin{aligned}\mathbf{F} &= \overline{\lambda \nabla T} = \lambda_p \overline{X_p \nabla T} + \lambda_f \overline{X_f \nabla T} \\ &= \lambda_f \overline{(X_p + X_f)\nabla T} + (\lambda_p - \lambda_f)\overline{X_p \nabla T} \\ &= \lambda_f \mathbf{G} + (\lambda_p - \lambda_f)\overline{X_p \nabla T} \qquad (5.23)\end{aligned}$$

Let us compute

$$\begin{aligned}\overline{X_p \nabla T} = \alpha_p \mathbf{q}_p &= \int f^{(1)}(\mathbf{z}_1)\overline{\mathbf{q}}^{(1)}(\mathbf{x}, t|\mathbf{z}_1) d\mathbf{z}_1 \\ &= \int\int f^{(2)}(\mathbf{z}_1, \mathbf{z}_2)\overline{\mathbf{q}}^{(2)}(\mathbf{x}, t|\mathbf{z}_1, \mathbf{z}_2) d\mathbf{z}_1 d\mathbf{z}_2 \qquad (5.24)\end{aligned}$$

**5.2.1. Conductivity-hierarchical method.** The hierarchy method is based on the assumption that the conditionally averaged fields are actually due to the spheres in an infinite uniform matrix. Thus, we assume that

$$\overline{\mathbf{q}}^{(1)}(\mathbf{x}, t|\mathbf{z}_1) = \mathbf{q}^{(1)}(\mathbf{x}, t|\mathbf{z}_1) \qquad (5.25)$$

and

$$\overline{\mathbf{q}}^{(2)}(\mathbf{x}, t|\mathbf{z}_1, \mathbf{z}_2) = \mathbf{q}^{(2)}(\mathbf{x}, t|\mathbf{z}_1, \mathbf{z}_2) \qquad (5.26)$$

*Order* $\alpha_p$. If we neglect the presence of the other spheres in the one-particle average, we have the temperature around a single sphere given by

$$T = T_p + 3\mathbf{G} \cdot \mathbf{x} \qquad (5.27)$$

inside the sphere, and

$$T = T_{f0} + \mathbf{G} \cdot \mathbf{x} + (T_{f0} + \mathbf{z}_1 \cdot \mathbf{G} - T_p)\frac{a}{r} + \mathbf{G} \cdot \mathbf{x}\frac{a^2}{r^2} \qquad (5.28)$$

outside the sphere.

Then the average temperature gradient inside the sphere is

$$\overline{X_p \nabla T} = 3\alpha_p \mathbf{G} \qquad (5.29)$$

This gives, to order $\alpha_p$,

$$\alpha_f \mathbf{q}_f = \lambda_f \mathbf{G} = \lambda_f \alpha_f \nabla T_f \qquad (5.30)$$

$$\alpha_p \mathbf{q}_p = 3\lambda_p \alpha_p \nabla T_f \qquad (5.31)$$

$$E_f = \alpha_p \lambda_f (T_p - T_f)\frac{3}{a} - 3\lambda_p \nabla T \cdot \nabla \alpha_p \qquad (5.32)$$

*Order* $\alpha_p^2$. The solution for the temperature for two spheres of radius $a$ is

$$T = \mathbf{G}\cdot\mathbf{x} + \sum_{m=0}^{1}\sum_{n=m}^{\infty} G_m d_{mn}\left(\frac{r_2}{a}\right)^{n+1} P_n^m(\cos\theta_1)\cos m\phi \tag{5.33}$$

inside the sphere, and

$$T = \mathbf{G}\cdot\mathbf{x} + \sum_{m=0}^{1}\sum_{n=m}^{\infty}$$

$$G_m\left\{g_{mn}^{(1)}\left(\frac{a}{r_1}\right)^{n+1} P_n^m(\cos\theta_1) + g_{mn}^{(2)}\left(\frac{a}{r_2}\right)^{n+1} P_n^m(\cos\theta_2)\right\}\cos m\phi \tag{5.34}$$

outside both spheres. Here $G_0$ and $G_1$ are the components of $\mathbf{G}$ parallel and perpendicular to the the line of centers of the two spheres and $r_i$ is the distance from the space point $\mathbf{x}$ to the center of sphere $i$. The coordinate system used here is identical to that used by Jeffrey.

The coefficients $d_{mn}$, $g_{mn}^{(1)}$, and $g_{mn}^{(2)}$ are found from the boundary conditions for continuity of temperature and heat flux at the surface of each sphere. The spherical harmonic identity

$$\left(\frac{a}{r_1}\right)^{n+1} P_n^m(\cos\theta_1) = \frac{1}{R}\sum_{s=m}^{\infty}\binom{n+s}{s+m}\left(\frac{r_{s-i}}{R}\right)^{s} P_s^m(\cos\theta_{s-i}) \tag{5.35}$$

The resulting equations for the coefficients are

$$d_{mn} = g_{mn}^{(i)} + \sum_{s=m}^{\infty}\binom{n+s}{n+m} g_{ms}^{3-i}\left(\frac{a}{R}\right)^{n+s+1} \tag{5.36}$$

and

$$\begin{aligned} r_\lambda d_{mn} + \left(1+\frac{1}{n}\right) g_{mn}^{(i)} + \sum_{s=m}^{\infty}\binom{n+s}{n+m} g_{ms}^{3-i}\left(\frac{a}{R}\right)^{n+s+1} \\ = (-1)^{(m-1)}(r_\lambda - 1) r_\lambda \delta_{1n} \end{aligned} \tag{5.37}$$

where $r_\lambda = \lambda_p/\lambda_f$, and

$$\beta = \frac{\lambda_p - \lambda_f}{\lambda_p + 2\lambda_f} = \frac{r_\lambda - 1}{r_\lambda + 2} \tag{5.38}$$

$$\beta_n = \frac{n(r_\lambda - 1)}{n r_\lambda + n + 1} \tag{5.39}$$

These equations can be solved by Taylor series [5].

In order to proceed further, we shall assume [4] that the nearest sphere is centered at the space point: $\mathbf{z}_1 = \mathbf{x}$. Then the average temperature gradient is given by

$$\overline{X_p \nabla T} = \int\limits_{|\mathbf{z}_1-\mathbf{x}|<a} \int f^{(2)}(\mathbf{z}_1,\mathbf{z}_2) \nabla T^{(2)}(\mathbf{x}|\mathbf{z}_1,\mathbf{z}_2)\, d\mathbf{z}_1 d\mathbf{z}_2$$

$$\approx \int\limits_{|\mathbf{z}_1-\mathbf{x}|<a} \int f^{(2)}(\mathbf{z}_1,\mathbf{z}_2) \nabla T(\mathbf{x}|\mathbf{z}_1,\mathbf{z}_2)\, d\mathbf{z}_1 d\mathbf{z}_2$$

$$\approx \int\limits_{|\mathbf{z}_1-\mathbf{x}|<a} \int f^{(2)}(\mathbf{z}_1,\mathbf{z}_2) \nabla T(\mathbf{z}_1,\mathbf{z}_2)\, d\mathbf{z}_1 d\mathbf{z}_2 \tag{5.40}$$

Note that if $f^{(2)}(\mathbf{z}_1,\mathbf{z}_2) = f^{(1)}(\mathbf{z}_1) f^{(1)}(\mathbf{z}_2)$, then the second integral does not converge. Jeffrey [4] has solved this problem by renormalization.

If we use the nearest neighbor pair distribution function, we have

$$\overline{X_p \nabla T} = \int\limits_{|\mathbf{z}_1-\mathbf{x}|<a} \int \hat{f}^{(2)}(\mathbf{x},\mathbf{z}_1,\mathbf{z}_2) \nabla T^{(2)}(\mathbf{x}|\mathbf{z}_1,\mathbf{z}_2)\, d\mathbf{z}_1 d\mathbf{z}_2$$

$$\approx \int\limits_{|\mathbf{z}_1-\mathbf{x}|<a} \int f^{(2)}(\mathbf{x},\mathbf{z}_1,\mathbf{z}_2) \nabla T(\mathbf{z}_1|\mathbf{z}_1,\mathbf{z}_2)\, d\mathbf{z}_1 d\mathbf{z}_2$$

$$\approx f^{(1)} \frac{4}{3}\pi a^3 \int\limits_{r_2=a}^{\infty} f^{(2)}(\mathbf{x},\mathbf{z}_1,\mathbf{z}_2) \nabla T(\mathbf{z}_1|\mathbf{z}_1,\mathbf{z}_2)\; 4\pi r_2^2 dr_2$$

$$\approx f^{(1)} \frac{4}{3}\pi a^3 e^{8\alpha_p} \int\limits_{r_2=a}^{\infty} \exp\left(-\frac{4}{3}\pi f_{(1)} r_2^3\right) \nabla T(\mathbf{z}_1|\mathbf{z}_1,\mathbf{z}_2)\; 4\pi r_2^2 dr_2 \tag{5.41}$$

The result has the form

$$\overline{\lambda \nabla T} = \lambda_{\text{eff}} \nabla \overline{T} \tag{5.42}$$

where

$$\lambda_{\text{eff}} = \lambda_f \Lambda(\alpha_p, r_\lambda) \tag{5.43}$$

The function $\Lambda$ is shown in Figures 1-3 for different values of the relative conductivity $r_\lambda$.

**5.3. Interfacial force and stresses.** A major problem in the study of suspensions is to evaluate the viscosity of a suspension in terms of the viscosity of the suspending fluid and of the properties of the particles. Even when the particles are spherical and rigid, attempts at such evaluations lead to theoretical difficulties.

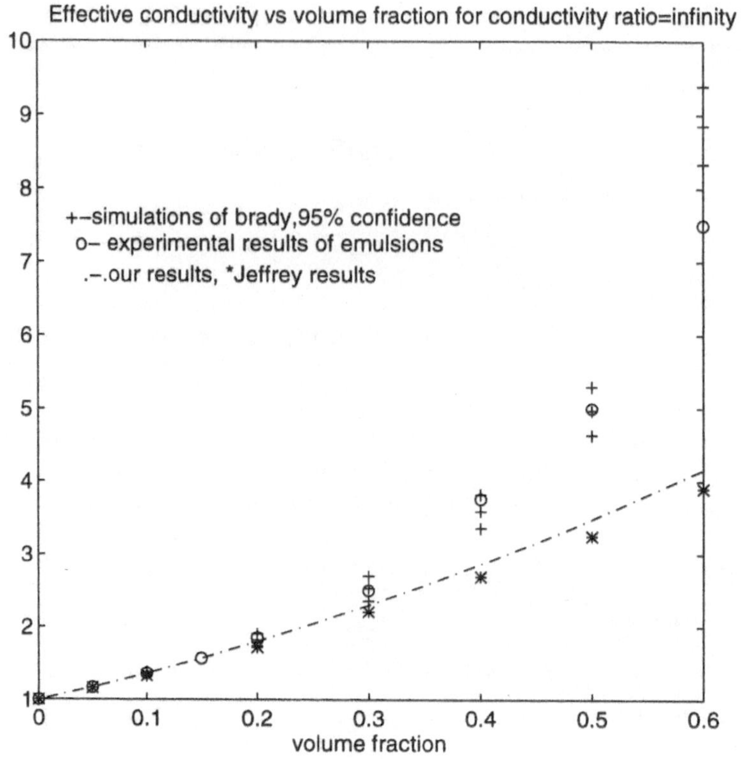


FIG. 1. *Relative Effective Conductivity,* $r_\lambda = \infty$

Several papers (notably Brenner [6], Batchelor [7], Hinch [2], and Lundgren [8]) have addressed the problem of slow viscous flow of a suspension of spheres, and of the flow of an inertia-less fluid through a fixed bed of spheres. These papers usually derive sedimentation velocities or effective viscosities for these mixtures. We shall adapt the arguments used in the literature to derive constitutive equations for the stresses and interfacial momentum transfer terms in (3.2) that are valid for slow viscous flow. We assume that the dispersed component consists of rigid elastic spheres of uniform composition, and that the continuous component is a fluid of constant density undergoing a motion that is sufficiently slow to be inertia-free.

*Constitutive Equations for Stress.* A fairly general expression for the stresses for slow flow is

$$\tau_p^m = 2\mu_{pp}\mathbf{D}_p + 2\mu_{pf}\mathbf{D}_f + E_{ppf}(\mathbf{v}_{pf}\nabla\alpha_p + \nabla\alpha_p\mathbf{v}_{pf})\,. \tag{5.44}$$

$$\tau_f^m = 2\mu_{fp}\mathbf{D}_p + 2\mu_{ff}\mathbf{D}_f + E_{fpf}(\mathbf{v}_{pf}\nabla\alpha_p + \nabla\alpha_p\mathbf{v}_{pf})\,. \tag{5.45}$$

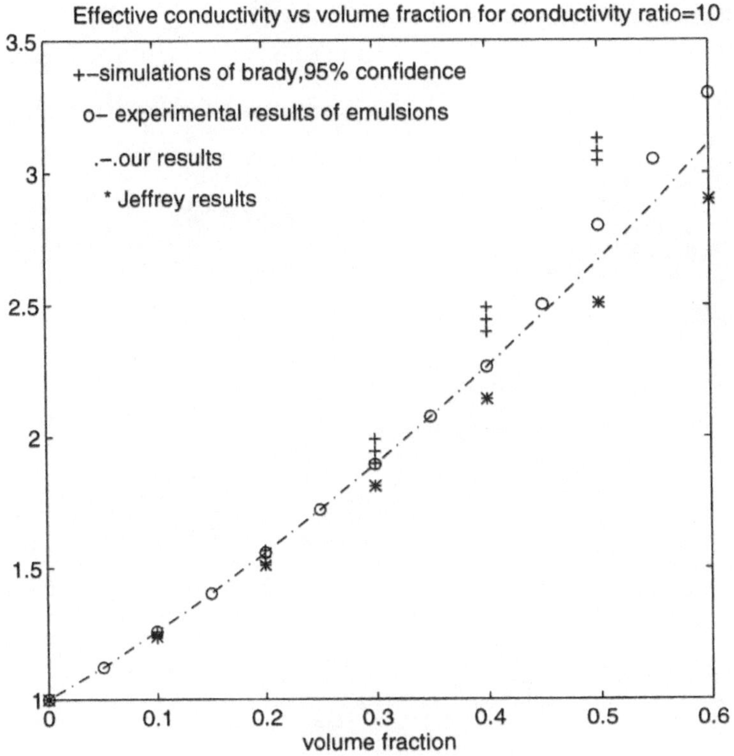


FIG. 2. *Relative Effective Conductivity,* $r_\lambda = 10.0$

where

$$\mathbf{D}_p = \frac{1}{2}\left[\nabla \mathbf{v}_p + (\nabla \mathbf{v}_p)^T\right] \tag{5.46}$$

$$\mathbf{D}_f = \frac{1}{2}\left[\nabla \mathbf{v}_f + (\nabla \mathbf{v}_f)^T\right] \tag{5.47}$$

are the rate of deformation tensors for the particles and the fluid, respectively, and

$$\mathbf{v}_{pf} = \mathbf{v}_p - \mathbf{v}_f \,. \tag{5.48}$$

A constitutive model for the mixture must specify the constitutive parameters $\mu_{pp}$, $\mu_{pp}$, $E_{ppf}$, and $E_{fpf}$.

Assume that the particles move with the fluid, so that $\mathbf{D}_p = \mathbf{D}_f$. Further assume that the mixture is uniform, so that $\nabla \alpha_p = 0$. Then adding Equations (5.44) and (5.45) gives

$$\alpha \tau_p^m + (1-\alpha)\tau_f^m = 2\left[\mu_{pp}(\alpha) + \mu_{pf}(\alpha) + \mu_{fp}(\alpha) + \mu_{ff}(\alpha)\right]\mathbf{D}_f \,. \tag{5.49}$$

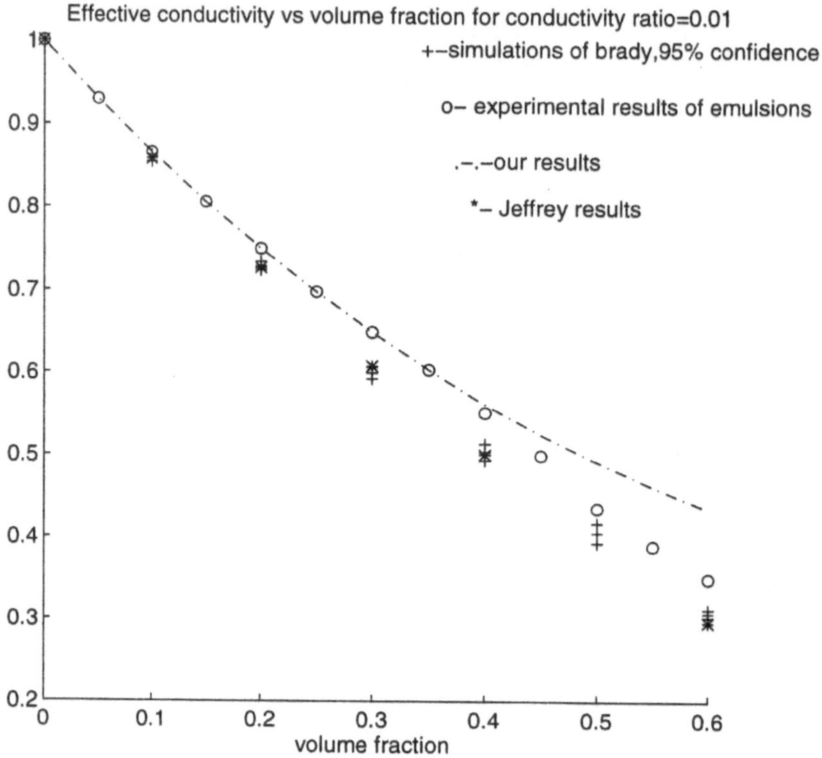


FIG. 3. *Relative Effective Conductivity,* $r_\lambda = 0.01$

Under these assumptions, the effective viscosity is given by

$$\mu_{\text{eff}} = \mu_{pp}(\alpha) + \mu_{pf}(\alpha) + \mu_{fp}(\alpha) + \mu_{ff}(\alpha) \; . \tag{5.50}$$

Thus, an expression for effective viscosity gives a relation between the various functions found in the stresses. We shall next discuss some of the existing expressions for effective viscosity.

*Effective Viscosity.* In an elegant classic paper, Einstein [10] evaluated the viscosity $\mu_{\text{eff}}$ of a linearly viscous fluid of viscosity $\mu_f$ containing a dilute suspension of small particles with volume fraction $\alpha_p$. The method consisted of calculating the energy dissipated by the flow around the particles, and associating that with the work done by the effective viscosity during shear of the mixture. Einstein's result is

$$\mu_{\text{eff}} = \mu_f \left(1 + \frac{5}{2}\alpha_p\right) \; . \tag{5.51}$$

Comparison of this equation with experimental data is not entirely gratifying. For small values of $\alpha_p$, that is, for dilute suspensions, the value of the viscosity and its increase with respect to $\alpha_p$ are correct. Deviation is soon noted, the trend being a vast under-estimation of the apparent viscosity as the solid volume fraction $\alpha_p$ increases. There is a large and interesting literature devoted to "correction" of the result of Einstein.

Most of the analysis is based on a calculation of Brenner [6]. There, the viscous dissipation rate due to two identical spheres approaching one another is computed. Frankel and Acrivos [11] show that this quantity is large compared to the dissipation rate associated with two spheres "passing" one another. For small distance of approach, the calculation is consistent with the lubrication approximation. They calculate the rate of dissipation of the suspension in an extensional flow, assuming that the suspended particles do not diffuse with respect to the fluid. That is set equal to the dissipation rate of an "effective" fluid, the result being

$$\frac{\mu_{\text{eff}}}{\mu_f} = C' \left\{ \frac{(\alpha_p/\alpha_m)^{\frac{1}{3}}}{1 - (\alpha_p/\alpha_m)^{\frac{1}{3}}} \right\}, \tag{5.52}$$

where $\alpha_m$ is a maximum packing fraction. It is to be determined experimentally. The functional form appears to be robust, the constant $C'$ being dependent on the details of the assumptions about the volume over which the dissipation is calculated. For example, the value

$$C' = \frac{9}{8} \tag{5.53}$$

results from assuming that a spherical shell is the "region of influence" around a given sphere, while

$$C' = \frac{3\pi}{16} \tag{5.54}$$

is the value computed from a cubic "region of influence". Frankel and Acrivos choose

$$C' = \frac{9}{8} \tag{5.55}$$

as the value that fits the data best. They note a number of other expressions in the literature, some empirical, some theoretical but with empirical parameters. None fit the data for large concentrations better than theirs does.

Furthermore, there is a generalization of the formula of Frankel and Acrivos [11] that agrees with Einstein's for small $\alpha_p$. It is given by Graham [12] as

$$\frac{\mu_{\text{eff}}}{\mu_f} = 1 + \frac{5}{2}\alpha_p + \frac{9}{2}\left[\frac{1}{\left(\frac{h}{a}\right)\left(2+\frac{h}{a}\right)\left(1+\frac{h}{a}\right)^2}\right], \tag{5.56}$$

where $a$ is the particle radius and $h$ is the interparticle spacing. For a simple cubic packing,

$$\frac{h}{a} = 2\frac{1-(\alpha_p/\alpha_m)^{\frac{1}{3}}}{(\alpha_p/\alpha_m)^{\frac{1}{3}}}, \tag{5.57}$$

The effective viscosity of a mixture of solid particles in a viscous fluid is dependent on the viscosity of the fluid and the volume fraction of solids in the fluid.

**5.3.1. Lundgren's stress.** A general expression for the stress in the fluid has been derived by Lundgren (see [8]).

The relation between the average stress and the exact stress in the fluid is given by

$$\alpha_f \mathbf{T}_f = -\alpha_f p_f \mathbf{I} + \mu_f \overline{X_f \left[\nabla \mathbf{v} + (\nabla \mathbf{v})^T\right]}. \tag{5.58}$$

Since the spheres are rigid, the deformation rate in them vanishes,

$$\left[\nabla \mathbf{v} + (\nabla \mathbf{v})^T\right] = 0. \tag{5.59}$$

Thus,

$$\overline{X_p \left[\nabla \mathbf{v} + (\nabla \mathbf{v})^T\right]} = 0. \tag{5.60}$$

Now

$$\overline{X_k[\nabla \mathbf{v}]} = \nabla \alpha_k \mathbf{v}_k - \overline{(\nabla X_k)\mathbf{v}}, \tag{5.61}$$

and

$$\overline{X_k[(\nabla \mathbf{v})^{\mathsf{T}}]} = (\nabla \alpha_k \mathbf{v}_k)^{\mathsf{T}} - \overline{\mathbf{v}(\nabla X_k)}. \tag{5.62}$$

If we add (5.58) and (5.60), use the no slip condition at the interface, which states that $\mathbf{v}$ is continuous across the interface between the particles and the fluid; and Equations (5.61) and (5.62) for $k = p$; and the result that $\nabla X_f = -\nabla X_p$, we have

$$\overline{X_f \left[\nabla \mathbf{v} + (\nabla \mathbf{v})^T\right]} = \nabla(\alpha_f \mathbf{v}_f + \alpha_p \mathbf{v}_p) + \left[\nabla(\alpha_f \mathbf{v}_f + \alpha_p \mathbf{v}_p)\right]^T. \tag{5.63}$$

Thus,

$$\alpha_f \mathbf{T}_f = -\alpha_f p_f \mathbf{I} + \mu_f \left\{\nabla(\alpha_f \mathbf{v}_f + \alpha_p \mathbf{v}_p) + \left[\nabla(\alpha_f \mathbf{v}_f + \alpha_p \mathbf{v}_p)\right]^T\right\}. \tag{5.64}$$

It is interesting that the assumption of no deformation inside the particles allows a derivation of the average *fluid* stress in terms of the "volumetric velocity" $\mathbf{v}_v = \alpha_f \mathbf{v}_f + \alpha_p \mathbf{v}_p$. We have

$$\nabla \cdot \mathbf{v}_v = 0, \tag{5.65}$$

$$\mu_f \nabla^2 \mathbf{v}_v - \nabla \alpha_f p_f + \mathbf{M}_f + \alpha_f \rho_f \mathbf{g} = 0. \tag{5.66}$$

Equations (5.65) and (5.66) do not form a complete model for any situation; at least a model for the interfacial force $\mathbf{M}_f$ is needed. Under most circumstances, the interaction will also involve the particle momentum equation. Thus, we must supply the constitutive equation for $\mathbf{T}_p$, as well as $\mathbf{M}_f$.

We wish to compute

$$\alpha_p \mathbf{T}_p = \overline{X_p \mathbf{T}} = \int\int f^{(1)}(\mathbf{z}_1)\mathbf{T}^{(1)}(\mathbf{x}, t|\mathbf{z}_1) d\mathbf{z}_1 \tag{5.67}$$

$$= \int\int f^{(2)}(\mathbf{z}_1, \mathbf{z}_2)\mathbf{T}^{(2)}(\mathbf{x}, t|\mathbf{z}_1, \mathbf{z}_2) d\mathbf{z}_1 d\mathbf{z}_2 \tag{5.68}$$

and

$$\mathbf{M}_f = \overline{\mathbf{T}\nabla X_f} = \int_{|\mathbf{z}_1 - \mathbf{x}| = a} f^{(1)}(\mathbf{z}_1)\mathbf{n} \cdot \mathbf{T}^{(1)}(\mathbf{x}, t|\mathbf{z}_1) d\mathbf{z}_1$$

$$= \int \int_{|\mathbf{z}_1 - \mathbf{x}| = a} f^{(2)}(\mathbf{z}_1, \mathbf{z}_2)\mathbf{n} \cdot \mathbf{T}^{(2)}(\mathbf{x}, t|\mathbf{z}_1, \mathbf{z}_2) d\mathbf{z}_1 d\mathbf{z}_2 \tag{5.69}$$

In order to proceed further, we again shall assume [1] that the nearest sphere is centered at the space point: $\mathbf{z}_1 = \mathbf{x}$. Then we have

$$\nabla \cdot (\mathbf{T}\mathbf{x}) = \mathbf{T} + \nabla \cdot (\mathbf{T})\mathbf{x} \tag{5.70}$$

However, we note that

$$\nabla \cdot \mathbf{T} = 0 \tag{5.71}$$

inside the particles. Then the average stress is given by

$$\alpha_p \mathbf{T}_p = \int_{|\mathbf{z}_1 - \mathbf{x}| = a} f^{(1)}(\mathbf{z}_1)\mathbf{n} \cdot \mathbf{T}^{(1)}(\mathbf{x}, t|\mathbf{z}_1)\mathbf{x} d\mathbf{z}_1$$

$$= \int \int_{|\mathbf{z}_1 - \mathbf{x}| = a} f^{(2)}(\mathbf{z}_1, \mathbf{z}_2)\mathbf{n} \cdot \mathbf{T}^{(2)}(\mathbf{x}, t|\mathbf{z}_1, \mathbf{z}_2)\mathbf{x} d\mathbf{z}_1 d\mathbf{z}_2 \tag{5.72}$$

**5.3.2. Stokes flow-hierarchical method.** If the suspension is dilute, the average stress conditioned on the presence of a sphere at $\mathbf{z}$, $\overline{\mathbf{T}}^{(1)}(\mathbf{x}, t|\mathbf{z}_1)$, is assumed to be the flow of the clear fluid around a single sphere in an unbounded fluid. Thus, we assume

$$\overline{\mathbf{T}}^{(1)}(\mathbf{x}, t|\mathbf{z}_1) = \mathbf{T}^{(1)}(\mathbf{x}, t|\mathbf{z}_1) \tag{5.73}$$

*Viscous Flow Around a Sphere.* Consider the flow of a fluid of viscosity $\mu_f$ around moving rigid spheres. The Stokes equations for the velocity field and the pressure are

$$\nabla \cdot \mathbf{v} = 0\,, \tag{5.74}$$

$$\nabla \cdot \mathbf{T} + \rho_f \mathbf{g} = 0\,, \tag{5.75}$$

where the stress is given by

$$\mathbf{T} = -p\mathbf{I} + \mu_f \left[\nabla\mathbf{v} + (\nabla\mathbf{v})^T\right]\,. \tag{5.76}$$

The solid material is assumed to be rigid.

*Order* $\alpha_p$. The equations of motion for the flow of the clear fluid around a single sphere in an unbounded fluid are

$$\nabla \cdot \mathbf{v}^{(1)} = 0\,, \tag{5.77}$$

$$0 = -\nabla p^{(1)} + \mu_f \nabla^2 \mathbf{v}^{(1)}\,. \tag{5.78}$$

We abbreviate $\mathbf{v}^{(1)}$ as $\mathbf{v}$.

Assume that the sphere is translating with respect to the fluid with velocity $\mathbf{v}_p$, and is rotating at angular velocity $\boldsymbol{\omega}$. Then

$$\mathbf{v} = \mathbf{v}_p + \mathbf{x} \times \boldsymbol{\omega} \quad \text{at} \quad r = a, \tag{5.79}$$

where $r = |\mathbf{x}|$. As $r \to \infty$ we require

$$\mathbf{v} \to \mathbf{v}_f + \mathbf{x} \cdot \nabla \mathbf{v}_f\,, \tag{5.80}$$

where $\mathbf{v}_f$ is the velocity of the fluid far from the sphere. It is also the velocity that would exist at the sphere if the sphere were not present.

We shall assume that no external torques are exerted on the sphere. Then, the rotation rates of the fluid and the sphere are the same.

$$\boldsymbol{\omega} = \nabla \times \mathbf{v}_f\,. \tag{5.81}$$

The solution to the fluid motion problem is in standard fluid mechanics references [13,14]. The form that we shall use is

$$p = p^t + p^s\,, \tag{5.82}$$

$$\mathbf{v} = \mathbf{v}_f + \mathbf{v}^t + \mathbf{v}^s\,, \tag{5.83}$$

where the translation part of the solution is given by

$$p^t = \frac{3\mu_f a}{2}(\mathbf{v}_p - \mathbf{v}_f) \cdot \frac{\mathbf{x}'}{|\mathbf{x}'|^3}\,, \tag{5.84}$$

$$\begin{aligned} \mathbf{v}^t &= \frac{3a}{4}(\mathbf{v}_p - \mathbf{v}_f) \cdot \left(\frac{\mathbf{x}'\mathbf{x}'}{|\mathbf{x}'|^3} + \frac{\mathbf{I}}{|\mathbf{x}'|}\right) \\ &\quad + \frac{a^3}{4}(\mathbf{v}_p - \mathbf{v}_f) \cdot \left(\frac{\mathbf{I}}{|\mathbf{x}'|^3} - \frac{3\mathbf{x}'\mathbf{x}'}{|\mathbf{x}'|^5}\right), \end{aligned} \tag{5.85}$$

and the linear shearing part is given by

$$p^s = \frac{5\mu_f a^5}{2}\mathbf{L}_f : \frac{\mathbf{x}'\mathbf{x}'\mathbf{x}'}{|\mathbf{x}'|^7}, \tag{5.86}$$

$$\begin{aligned} \mathbf{v}^s &= -\frac{5a^3\mathbf{L}_f}{2} : \left(\frac{\mathbf{x}'\mathbf{x}'\mathbf{x}'}{|\mathbf{x}'|^5}\right) \\ &\quad + \mathbf{L}_f : \left(\frac{5a^5\mathbf{x}'\mathbf{x}'\mathbf{x}'}{2|\mathbf{x}'|^7} - \frac{a^5(\mathbf{x}'\mathbf{I} + \mathbf{I}\mathbf{x}')}{2|\mathbf{x}'|^5}\right). \end{aligned} \tag{5.87}$$

where $\mathbf{L}_f$ is the rate of strain tensor in the fluid far from the particle, and $\mathbf{x}'$ is the vector from the center of the sphere to the space point $\mathbf{x}$.

If we perform the average, we have the average force density at point $\mathbf{x}$ is

$$\begin{aligned} \mathbf{M}_p &= \frac{1}{\frac{4}{3}\pi R^3} \int_{|\mathbf{x}'|=a} \mathbf{n} \cdot \mathbf{T}^{(1)}(\mathbf{x}, t|\mathbf{x} - \mathbf{x}')\, d\Omega_{\mathbf{x}'}, \\ &= \frac{9}{2}\frac{\alpha_p}{a^2}(\mathbf{v}_f - \mathbf{v}_p) + \frac{5}{2}\mu_f \mathbf{L}_f \cdot \nabla\alpha_p . \end{aligned} \tag{5.88}$$

The first part of this expression is the classical Stokes drag [15].

The average stress inside the particles can be found from (5.67) to be

$$\mathbf{T}_p = \frac{5}{2}\mu_f \mathbf{L}_f . \tag{5.89}$$

We note that the calculation of Lundgren [8] of the fluid stress, to within neglecting terms of $O(\alpha_p)$ gives

$$\mathbf{T}_f = \mu_f \mathbf{L}_f . \tag{5.90}$$

*Order* $\alpha_p^2$. The solution for the velocity field in an unbounded fluid with two spheres with no shearing far from the spheres is given by Jeffrey and Onishi [16]. Hurwitz [17] gives an expression that does contain the liquid shearing. If we are interested in corrections for small concentrations, we can assume that the second sphere is far from the first sphere. In that case, the velocity field can be found by considering the perturbation of the velocity in the neighborhood of the sphere at $\mathbf{z}_1$ due to the sphere at $\mathbf{z}_2$. The velocity in the unbounded fluid due to the sphere at $\mathbf{z}_1$ is given by (5.85) and (5.87). This induces a perturbation velocity at $\mathbf{z}_2$ of

$$\mathbf{v}^{(1)}(\mathbf{z}_2) \approx \frac{3a}{4}(\mathbf{v}_p - \mathbf{v}_f) \cdot \left(\frac{\mathbf{z}_2'\mathbf{z}_2'}{|\mathbf{z}_2'|^3} + \frac{\mathbf{I}}{|\mathbf{z}_2'|}\right) - \frac{5a^3\mathbf{L}_f}{2} : \left(\frac{\mathbf{z}_2'\mathbf{z}_2'\mathbf{z}_2'}{|\mathbf{z}_2'|^5}\right) \tag{5.91}$$

to order $1/|\mathbf{z}_2'|^3$ where

$$\mathbf{z}_2' = \mathbf{z}_2 - \mathbf{z}_1 . \tag{5.92}$$

By a result of Faxén [13], there results a force on the sphere at $\mathbf{z}_2$ given by

$$\mathbf{f}_2'(\mathbf{z}_2') = 6\pi\mu_f a\left(1 + \frac{a^2}{6}\nabla^2\right)\mathbf{v}^{(1)}(\mathbf{z}_2), \tag{5.93}$$

to order $1/|\mathbf{z}_2'|^3$. This force results in a velocity field at $\mathbf{x}$ of

$$\mathbf{v}^{(2)}(\mathbf{x}) \approx \mathbf{v}^{(1)} + \mathbf{v}' \tag{5.94}$$

where

$$\mathbf{v}'(\mathbf{x}) = \frac{\mathbf{f}_2'(\mathbf{z}_2')}{8\pi\mu_f}\cdot\left(\frac{\mathbf{x}''\mathbf{x}''}{|\mathbf{x}''|^3} + \frac{\mathbf{I}}{|\mathbf{x}''|}\right) \tag{5.95}$$

to order $1/|\mathbf{x}''|^2$, where $\mathbf{x}'' = \mathbf{x} - \mathbf{z}_2$. Note that since $\mathbf{f}_2'$ is $O(1/|\mathbf{z}_2|^2)$ and the Stokeslet is $O(1/|\mathbf{z}_2|)$, the stress due to $\mathbf{v}'$ is $O(1/|\mathbf{z}_2|^3)$, and the integral over $\mathbf{z}_2$ must be treated carefully. The "offending" part comes from the Stokeslet in (5.95), and can be treated in several different ways.

*Renormalization.* One way to treat the possibility of non-convergence is to "renormalize" the problem. This method consists of recognizing that the origin of the singular term is due to a drag force on the sphere at $\mathbf{z}_2$, and should be treated with the other drag terms. Batchelor [1], and others using the hierarchical method, treat this problem by adding and subtracting the singular part of the stress from the stress $\mathbf{T}$ in the interfacial force density, and evaluating the singular part of the stress contribution by a distributed force density. As we see from (5.95), the singular part is the velocity caused by a point force

$$\mathbf{f}' = \frac{9\pi\mu_f}{2}(\mathbf{v}_p - \mathbf{v}_f)\delta(\mathbf{x} - \mathbf{z}_2), \tag{5.96}$$

resulting in a Stokeslet behavior:

$$\mathbf{v}_s = \frac{\mathbf{f}'}{8\pi}\cdot\left(\frac{\mathbf{x}''\mathbf{x}''}{|\mathbf{x}''|^3} + \frac{\mathbf{I}}{|\mathbf{x}''|}\right). \tag{5.97}$$

If this point force is multiplied by $f^{(1)}(\mathbf{z}_2)$ and integrated over all space, the result is

$$\overline{\mathbf{T}_p \cdot \nabla X_f} = \frac{9}{2a^2}\alpha_p(\mathbf{v}_p - \mathbf{v}_f), \tag{5.98}$$

the Stokes drag on a single sphere per unit mixture volume.

Thus, we wish to solve the problem

$$\nabla\cdot\mathbf{v} = 0, \tag{5.99}$$

$$\mu_f\nabla^2\mathbf{v} - \nabla p - \alpha_p S\mathbf{v} = 0. \tag{5.100}$$

with the boundary condition that the velocity must approach

$$\mathbf{v} = \mathbf{x} \cdot \mathbf{L}_f . \tag{5.101}$$

The average stress inside the particle can be found from (5.68) to be

$$\mathbf{T}_p = \left(5 + \frac{515}{32}\alpha_p\right) \mu_f \mathbf{L}_f . \tag{5.102}$$

The interfacial force density can be obtained from (5.18). It is

$$\mathbf{M}_p = \frac{9\alpha_p \mu_f}{2a}\left(1 - \frac{845}{128}\alpha_p\right)(\mathbf{v}_f - \mathbf{v}_p) . \tag{5.103}$$

We note that the solution given by Equations (5.85) and (5.87) cannot be correct at point $\mathbf{x}$ if some other sphere is closer to $\mathbf{x} = \mathbf{z}_1$ than the one at $\mathbf{z}_2$. Note further that if we replace $f^{(2)}(\mathbf{z}_1, \mathbf{z}_2)$ by $\hat{f}^{(2)}(\mathbf{z}_1, \mathbf{z}_2, \mathbf{x})$, the nearest neighbor pair distribution function, then the integral to calculate $\mathbf{T}_p$ converges because of the exponential decay in $\hat{f}^{(2)}$.

**5.3.3. Effective medium theory.** The methods in the previous subsection assumed that the single sphere or the pair of spheres experience conditions due to the flow of a clear fluid around then; i. e.,

$$\overline{\phi}^{(i)}(\mathbf{x}|\mathbf{z}_1, \ldots) \approx \phi^{(1)}(\mathbf{x}|\mathbf{z}_1, \ldots). \tag{5.104}$$

It is equally sensible to assume that the single sphere or pair of spheres experience conditions due to an *effective medium,* that is, one where the mechanics is given by equations with terms representing effective properties. The original idea for this method is due to Brinkman [18]. It has been used extensively in studies of electromagnetic and acoustic wave propagation in inhomogeneous media. There are difficulties with this method. First, the method is predicated on the replacement of the effects of the dynamics of *two* components with the dynamics of one effective medium. There is no unique way to do this. At least two separate ways are obvious, holding the particles fixed, and letting the particles move with the fluid. The second difficulty is in the treatment of the effect of the field near the fixed particle. This is related to the difficulty with what to use for $\alpha_p^{(1)}$. We note that we know of no exact solutions of the equations of slow viscous flow around a sphere when the viscosity is variable. Therefore, we expect that the effective medium equations are not soluble if we allow nonuniform particle density. Clearly, however, the fixed particle is actually in contact with the clear fluid almost all of the time, over almost all of its surface area. Lundgren [8] uses the effective medium right up to the sphere. This is not entirely accurate. Recent researchers [17,19] have instead introduced a "blending" function accounting for the presence of the clear fluid near the sphere, and gradually becoming the effective medium far from the sphere.

However, they solve the conditionally averaged equations without considering the nonuniform volume fraction. Itoh [19] uses a linear distribution.

If we now wish to calculate the interfacial force density

$$\mathbf{M}_p = \int_{|\mathbf{z}-\mathbf{x}|=a} f^{(1)}(\mathbf{z},t)\,\overline{\mathbf{T}}^{(1)}(\mathbf{x},t|\mathbf{z})\cdot\mathbf{n}\,d\Omega\,, \tag{5.105}$$

we must derive an expression for $\overline{\mathbf{T}}^{(1)}(\mathbf{x},t|\mathbf{z})$, the averaged stress at $\mathbf{x}$ given that there is a sphere centered at $\mathbf{z}$. It is convenient to start with (5.4) and (5.16) for the dynamics of the fluid with one sphere held fixed. These can be written as

$$\nabla\cdot\overline{\mathbf{v}}_v^{(1)} = 0\,, \tag{5.106}$$

$$\mu_f\nabla^2\overline{\mathbf{v}}_v^{(1)} + \mathbf{M}_p^{(1)} - \nabla\alpha_f^{(1)}\overline{p}_f^{(1)} = 0\,, \tag{5.107}$$

where

$$\overline{\mathbf{v}}_v^{(1)} = \alpha_p^{(1)}\overline{\mathbf{v}}^{x(1)} + \alpha_f^{(1)}\overline{\mathbf{v}}_f^{x(1)} \tag{5.108}$$

is the volumetric velocity. We assume that there is no body force on the fluid component, so that $\mathbf{g}_f = 0$. We shall use the body force on the particles to justify certain special assumptions about the particle velocity. We shall use the assumption that $\alpha_p^{(1)} = \alpha_p =$ constant. Furthermore, we shall abbreviate $\overline{\mathbf{v}}_v^{(1)}$ by $\mathbf{v}_v$. The remainder of the procedure for obtaining expressions for the effective viscosity and the drag is to assume a form for $\mathbf{M}_p$ in terms of $\alpha_p$, $\mathbf{v}_p$, and $\mathbf{v}_f$, *assume the same form* for $\mathbf{M}_p^{(1)}$, solve the resulting boundary value problem for the flow around the sphere centered at $\mathbf{z}$, and use the stress to compute $\mathbf{M}_p$ from (5.105). The resulting equation will contain coefficients from the form for $\mathbf{M}_p$. The strategy is to get one equation for one unknown coefficient at a time. Then it is simple to solve for the coefficient.

For slow viscous flow with constant particle density, the form

$$\mathbf{M}_p = \alpha_p\overline{S}^e(\alpha_p)(\mathbf{v}_f - \mathbf{v}_p) - 2\overline{\mu}^{if}(\alpha_p)\,\mathbf{D}_f\cdot\nabla\alpha_p - 2\overline{\mu}^{ip}(\alpha_p)\,\mathbf{D}_p\cdot\nabla\alpha_p \tag{5.109}$$

appears to be adequate. Also, it is less obvious at this point that we need an expression for $\mathbf{T}_p$; however, we give one here. We assume that

$$\mathbf{T}_p = -p_f\mathbf{I} + 2\overline{\mu}_p^{ep}(\alpha_p)\mathbf{D}_p + 2\overline{\mu}_p^{ef}(\alpha_p)\mathbf{D}_f\,. \tag{5.110}$$

*Effective Viscosity.* For this calculation of the effective viscosity, we assume that a uniform distribution of particles move with the fluid, so that $\nabla\alpha_p = 0$ and $\mathbf{v}_p = \mathbf{v}_f = \mathbf{v}$. Then the continuity equation (5.8) gives

$$\nabla\cdot\mathbf{v} = 0\,, \tag{5.111}$$

and the momentum equation (5.9) for the fluid becomes

$$\nabla^2[\overline{\mu}^e(\alpha_p)\mathbf{v}] - \nabla p = 0\,, \tag{5.112}$$

where

$$\overline{\mu}^e(\alpha_p) = \mu_f + \overline{\mu}^{ep} + \overline{\mu}^{ef}\,, \tag{5.113}$$

and

$$p = \alpha_f p_f\,. \tag{5.114}$$

The boundary conditions that we use are

$$\mathbf{v} = 0 \quad \text{on} \quad |\mathbf{x} - \mathbf{z}| = a\,, \tag{5.115}$$

and

$$\mathbf{v} \to \mathbf{v}_f \quad \text{as} \quad |\mathbf{x} - \mathbf{z}| \to \infty\,. \tag{5.116}$$

We again take $\mathbf{v}_f = \mathbf{v}_{f0} + \mathbf{L}_f \cdot (\mathbf{x} - \mathbf{z})$, where $\mathbf{L}_f$ is a constant tensor with $\operatorname{tr}\mathbf{L}_f = 0$.

Equation (5.112) is just the equation for Stokes flow, with an effective viscosity. The solution to this problem is given by (5.84), (5.85), (5.86) and (5.87). Thus, proceeding as before, we find that the average stress in the fixed particle is given by

$$\mathbf{T}_p = \frac{5}{2}\mu^e(\alpha_p)\mathbf{L}_f\,. \tag{5.117}$$

Thus, we must have

$$\frac{5}{2}\overline{\mu}^e = \overline{\mu}^{ep} + \overline{\mu}^{ef}\,. \tag{5.118}$$

The equation defining the fluid stress is (5.64). Thus, the average stress in the fluid is

$$(1 - \alpha_p)\mathbf{T}_f = \mu_f \mathbf{L}_f\,. \tag{5.119}$$

The effective stress is the mixture stress,

$$\begin{aligned} \overline{\mathbf{T}} &= (1 - \alpha_p)\mathbf{T}_f + \alpha_p \mathbf{T}_p\,, \\ &= \mu_f \mathbf{L}_f + \frac{5}{2}\alpha_p \overline{\mu}^e(\alpha_p)\mathbf{L}_f\,, \\ &= \overline{\mu}^e(\alpha_p)\mathbf{L}_f\,. \end{aligned} \tag{5.120}$$

Solving for $\overline{\mu}^e(\alpha_p)$ gives

$$\overline{\mu}^e(\alpha_p) = \frac{\mu_f}{1 - \frac{5}{2}\alpha_p}\,, \tag{5.121}$$

and, from (5.117)

$$\overline{\mu}_p^{ep}(\alpha_p) + \overline{\mu}_p^{ef}(\alpha_p) = \frac{\frac{5}{2}\mu_f}{1 - \frac{5}{2}\alpha_p} . \tag{5.122}$$

Lundgren notes that this expression becomes infinite at $\alpha_p = 0.4$. He claims that at such concentrations, the assumption that a suspension behaves as a Newtonian fluid is probably not valid. Note that for small $\alpha_p$, the effective viscosity $\mu^e(\alpha_p)$ can be expanded in a Taylor series in $\alpha_p$. The result is

$$\overline{\mu}^e(\alpha_p) \approx \mu_f \left(1 + \frac{5}{2}\alpha_p + \frac{25}{4}\alpha_p^2 + O(\alpha_p^3)\right) . \tag{5.123}$$

When terms of $O(\alpha_p^2)$ are neglected, this reduces to Einstein's [10] result. The result for $\overline{\mu}^e(\alpha_p)$ agrees well with experimental data; perhaps better than should be expected.

*Drag.* In order to calculate the drag by effective medium theory, consider the problem given by (5.111–5.112), with boundary conditions

$$\mathbf{v} = \mathbf{v}_p \quad \text{on} \quad |\mathbf{x} - \mathbf{z}| = a \tag{5.124}$$

and

$$\mathbf{v} \to \mathbf{v}_f \quad \text{as} \quad |\mathbf{x} - \mathbf{z}| \to \infty . \tag{5.125}$$

We now take

$$\mathbf{v}_f = \mathbf{v}_{f0} = V_0 \mathbf{e}_z , \tag{5.126}$$

where $V_0$ is a constant. Without loss of generality, we can take $\mathbf{v}_p = 0$. For the calculation of the effective drag coefficient, we assume that $\alpha_p$ is constant. Thus, we must solve the problem

$$\nabla \cdot \mathbf{v} = 0 , \tag{5.127}$$

$$(1 - \alpha_p)\mu_f(\alpha_p)\nabla^2 \mathbf{v} - \nabla p - \alpha_p \overline{S}^e(\alpha_p)\mathbf{v} = 0 . \tag{5.128}$$

with the boundary condition that the velocity must approach

$$\mathbf{v}_{f0} = V_0 \mathbf{e}_z . \tag{5.129}$$

We see that $\nabla^2 p = 0$. Itoh [19] gives the solution for the velocity and pressure as follows. First, assume that the pressure is of the form

$$p = (1 - \alpha_p)\mu_f(\alpha_p) V_0 \chi(r) \cos\theta, \tag{5.130}$$

where

$$r = |\mathbf{x} - \mathbf{z}| ,$$

and $\theta$ is the angle between $\mathbf{v}_{f0}$ and $\mathbf{x}-\mathbf{z}$. Since the pressure is harmonic, the function $\chi$ must satisfy Laplace's equation in spherical coordinates, so that its $r$ dependence is given by

$$\chi_{rr}+\frac{2}{r}\,\chi_r-\frac{2}{r^2}\,\chi=0\,, \tag{5.131}$$

with solution

$$\chi=C_0 r+\frac{C}{r^2}\,, \tag{5.132}$$

where $C_0$ and $C$ are constants.

Writing

$$\mathbf{v}_{f0}=V_0\mathbf{e}_z=V_0(\cos\theta\;\mathbf{e}_r-\sin\theta\;\mathbf{e}_\theta)$$

suggests that the solution for the velocity should be of the form

$$\mathbf{v}_f=V_0\left[\phi\cos\theta\;\mathbf{e}_r-\left\{\phi+\frac{r}{2}\,\phi_r\right\}\sin\theta\;\mathbf{e}_\theta\right]. \tag{5.133}$$

Substituting (5.133) into (5.128) results in the following equation for $\phi$:

$$\phi_{rr}+\frac{4}{r}\,\phi_r-\hat{S}\phi-\chi_r=0\,, \tag{5.134}$$

where $\hat{S}=\alpha_p\overline{S}^e(\alpha_p)/(1-\alpha_p)\mu_f(\alpha_p)$. The boundary conditions for $\phi$ are

$$\phi(a)=0\,, \tag{5.135}$$

$$\phi_r(a)=0\,, \tag{5.136}$$

$$\phi\rightarrow 1\quad\text{as}\quad r\rightarrow\infty\,. \tag{5.137}$$

The solution can be written as

$$\phi=-\frac{C_0}{\hat{S}}+\frac{2C}{r^3\hat{S}}+C_2\frac{1+\sqrt{\hat{S}}r}{r^3}e^{-\sqrt{\hat{S}}r}\,, \tag{5.138}$$

where $C_2$ is a constant of integration. There is a second solution of the homogeneous equation that grows like $\exp(\hat{S}^{\frac{1}{2}}r)$. It has been left out. The boundary conditions give

$$C_0=-\hat{S}\,, \tag{5.139}$$

$$C=-\frac{3a+3a^2\sqrt{\hat{S}}+a^3\hat{S}}{2}\,, \tag{5.140}$$

$$C_2 = \frac{3ae^{a\sqrt{\hat{s}}}}{\hat{S}}. \tag{5.141}$$

In terms of $\phi$ and $\chi$, the stress is given by

$$\begin{aligned} \mathbf{T} = \mu_f(\alpha_p)V_0\{&-\chi\cos\theta\mathbf{I} + (1-\alpha_p)2\phi_r\cos\theta\mathbf{e}_r\mathbf{e}_r \\ &-\frac{1}{2}\left(r\phi_{rr} + \frac{2}{r}\phi\right)\sin\theta(\mathbf{e}_r\mathbf{e}_\theta + \mathbf{e}_\theta\mathbf{e}_r) \\ &+\frac{1}{r}\ \phi\cos\theta(\mathbf{e}_\theta\mathbf{e}_\theta + \mathbf{e}_\omega\mathbf{e}_\omega)\}\,, \end{aligned} \tag{5.142}$$

where $\mathbf{e}_r$, $\mathbf{e}_\theta$, and $\mathbf{e}_\omega$ are the unit vectors in spherical coordinates. Taking the inner product of the stress with $\mathbf{n}$, and integrating over $r = a$ gives

$$-\overline{\mathbf{T}\cdot\nabla X_p} \quad = \quad -2\pi a^2(1-\alpha_p)\mu_f(\alpha_p)V_0\left[\phi_{rr}\big|_{r=a} + \chi(a)\right]\mathbf{e}_z\,.$$

Substituting for $\chi$ and $\phi$, and setting this equal to $\alpha_p\overline{S}^e(\alpha_p)V_0\mathbf{e}_z$, and solving for $\overline{S}^e$ gives

$$\overline{S}^e(\alpha_p) = \frac{9}{2a^2}\frac{1+\frac{3}{4}\left[3\alpha_p + \left(8\alpha_p + 9\alpha_p^2\right)^{\frac{1}{2}}\right]}{\left(1-\frac{3}{2}\alpha_p\right)^2}. \tag{5.143}$$

**6. Conclusion.** Some results have been presented for the flow and heat conduction of a medium consisting of a random array of spheres in a uniform matrix. The Constitutive equations for heat exchange, heat flux, viscosity, and interfacial force density are given for various assumptions about the averaging process. Most of these constitutive equations are given elsewhere in the literature; however, we compare some of these results with those obtained by using the nearest neighbor pair correlation function.

The nearest neighbor pair correlation function is derived in Section 4, and is used herein to derive the conductivity of a medium consisting of a random array of uniform spheres of conductivity $\lambda_p$ in a matrix of conductivity $\lambda_f$. The results of this calculation, for various values of $\alpha_p$ and $r_\lambda = \lambda_p/\lambda_f$ are shown in Figures 1-3. Also shown there are several other models, including the numerical calculations of Bonnecaze and Brady [20]. In addition, experimental measurements are also shown in these figures. Note that the calculation using the nearest neighbor pair correlation function does very well for a surprising range of volume fraction $\alpha_p$.

A version of effective media theory used for conductivity calculations is called Debye screening, which accounts for intervening particles by including a term in the effective medium equation representing the mean field effect of the intervening particles. This is very different from the nearest neighbor pair correlation function method, where the presence of intervening particles is treated by using a probability density function that decreases the effect of far away particles by assigning a low probability to the contribution of the field at the given space point.

## REFERENCES


[1] G.K. Batchelor, The Stress System in a Suspension of Force-Free Particles, Journal of Fluid Mechanics, 41, 545–570 (1970).

[2] E.J. Hinch, An Averaged Equation Approach to Particle Interactions in a Fluid Suspension, Journal of Fluid Mechanics, 83, 695–720 (1977).

[3] J.C. Maxwell, Electricity and Magnetism, Clarendon Press, Oxford (1873).

[4] D.J. Jeffrey, Conduction through a Random Suspension of Spheres, Proc. Royal Soc., London A 335, 355–367 (1973).

[5] D.K. Ross, The Potential due to Two Point Charges Each at the Centre of a Spherical Cavity and Embedded in a Dielectric Medium Aust. J. Phys., 21, 817–822 (1968).

[6] H. Brenner, The Slow Motion of a Sphere Through a Viscous Fluid Toward a Plane Surface, Chemical Engineering Science, 16, 242–251 (1961).

[7] G.K. Batchelor and J.T. Green, The Determination of the Bulk Stress in a Suspension of Spherical Particles to Order $c^2$, J. Fluid Mech., 56, 401-427, (1972).

[8] T.S. Lundgren, Slow Flow Through Stationary Random Beds and Suspensions of Spheres, J. Fluid Mech., 238, 579–598, (1972).

[9] D.D. Joseph and T.S. Lundgren, Ensemble Averaged and Mixture Theory Equations, International Journal of Multiphase Flow, 16, 35–42 (1990).

[10] A. Einstein, Eine neue Bestimmung der Moleküldimensionen, Annalen der Physik, 19, 289–306 (1906).

[11] N.A. Frankel and A. Acrivos, On the Viscosity of a Concentrated Suspension of Solid Spheres, Chemical Engineering Science, 22, 847–853 (1967).

[12] A.L. Graham, On the Viscosity of Suspensions of Solid Spheres, Applied Scientific Research, 37, 275–286 (1981).

[13] J. Happel and H. Brenner, *Low Reynolds Number Hydrodynamics*, Noordhoff, Leyden (1973).

[14] H. Lamb, Hydrodynamics, Cambridge University Press (1932).

[15] G.B. Stokes, On the Effect of the Internal Friction of Fluids on the Motion of Pendulums, Trans. Cambridge Phil. Soc., 9, p8; also in Mathematical and Physical Papers, Vol. 3, Johnson Reprint Corporation, New York, (1966).

[16] D.J. Jeffrey and Y. Onishi, Calculation of the Resistance and Mobility Functions for Two Unequal Rigid Spheres in Low Reynolds Number Flow, Journal of Fluid Mechanics, 139, 261–290 (1984).

[17] M.F. Hurwitz, Hydrodynamic Interactions of Many Rigid Particles, Ph. D. Thesis, Cornell University, Ithaca, NY (1996).

[18] H.C. Brinkman, A Calculation of the Viscous Force Exerted by a Flowing Fluid on a Dense Swarm of Particles, Applied Scientific Research, A1, 27–34 (1947).

[19] S. Itoh, The Permeability of a Random Array of Identical Rigid Spheres, Journal of the Physical Society of Japan, 52, 2379–2388 (1983).

[20] R.T. Bonnecaze and J.F. Brady, The Effective Conductivity of Random Suspensions of Spherical Particles, Royal Soc., London A 432, 445–465 (1991).

# BIFURCATION WITH SYMMETRY IN MULTI-PHASE FLOWS

MANFRED F. GÖZ*

**Abstract.** A common approach to describe particulate flows consists in the use of (volume or ensemble) averaged equations of motion. These are coupled sets of compressible Navier-Stokes equations representing a typical wave-hierarchy problem with dissipation and expressing another hierarchy of symmetry-breaking instabilities. The general features of these hierarchies, their connection, and the details of the first two stages of instabilities are discussed.

**1. Introduction.** Particulate flows are certainly more common in natural (geo- and biophysical phenomena) and industrial (mechanical or chemical processing of particles) processes than the widely investigated pure single-phase flows, but their understanding is hindered by two interrelated basic problems, a descriptional one and one of analysis. In general, a dispersed two-phase flow consists of a gas or liquid as the continuous fluid phase and a collocation of particles, liquid drops or gas bubbles as the discrete phase [1]. It is impossible to solve the flow problem analytically by taking account of each member of the discrete phase individually, and also the numerical approach is restricted to relatively few particulate elements. This has led to numerous efforts to derive averaged equations of motion for the dispersed phase, but then the problems of proper closure or constitutive relations for the dispersed phase and the correct form of the interaction forces between the two phases arise [2]. Yet another problem is the determination of solutions of these equations, so that the modelling assumptions can be checked against experiments. It is not surprising that this requires a good deal of numerical and analytical effort due to the complicated nature and the mere number of the equations.

My approach is based upon flow field equations describing two interacting and interpenetrating continua, which are applicable to, e.g., suspensions of particles in fluidized beds or gas bubbles rising in a liquid column. The constitutive and interaction relations need not necessarily be specified, but for the sake of convenience and clarity certain restrictions are imposed. Another restriction is that I will look mostly for solutions periodic in space and time. Utilizing the theory of bifurcation with symmetry [3], a general qualitative analysis of the model equations can be performed based on the symmetries of the base state and the information (or assumptions) about the nature of successive instabilities. The painting of the broad picture is complemented by its description as a wave-hierarchy problem. Then I will zoom into the details of the apparently most important first two stages of instabilities, the first one leading from the uniform base state to one- or two-dimensional solutions, the second one leading from the one- to a two-

* Princeton University, Department of Chemical Engineering, Princeton, NJ 08544.

dimensional solution. All of these solutions represent vertically travelling waves, the last one being reminiscent of "bubbles" in fluidized beds. The analysis of gas-liquid flows yields similar results.

**2. Model equations and base state.** The mass and momentum balance equations under investigation may be written in the following dimensionless form [4-6]:

$$
\begin{aligned}
&(2.1) & -\partial_t \phi + \operatorname{div}\,[(1-\phi)\,\mathbf{v}] &= 0\,, && \text{(particulate phase)}\\
&(2.2) & \partial_t \phi + \operatorname{div}\,(\phi \mathbf{u}) &= 0\,, && \text{(continuous phase)}\\
&(2.3) & F(1-\phi)(\partial_t \mathbf{v} + \mathbf{v}\cdot\nabla\mathbf{v}) &= -(1-\phi)\mathbf{k} + B(\phi)(\mathbf{u}-\mathbf{v}) - FG(\phi)\nabla\phi \\
& & &\quad -(1-\phi)\nabla p + \mu(\Delta + \kappa\nabla\nabla\cdot)\mathbf{v}\,, \\
&(2.4) & F\delta\phi(\partial_t \mathbf{u} + \mathbf{u}\cdot\nabla\mathbf{u}) &= -\delta\phi\mathbf{k} - B(\phi)(\mathbf{u}-\mathbf{v}) - \phi\nabla p \\
& & &\quad +\nu\mu(\Delta + \bar{\kappa}\nabla\nabla\cdot)\mathbf{u}\,.
\end{aligned}
$$

Here, $\phi$ denotes the volume fraction of the continuous phase ("voidage" in the case of fluidized beds), $\mathbf{u}$ and $\mathbf{v}$ are the local velocities of the continuous and particulate phase, respectively, $p$ is an effective fluid pressure, and $G(\phi)\nabla\phi$ with $G(\phi) < 0$, represents a pressure contribution due to interactions within the particulate phase. The unit vector $\mathbf{k}$ points into the $x$-direction, which means against gravity; the horizontal coordinate is denoted by $y$. The equations have been made dimensionless by employing an average particle or gas bubble radius $r$ as length scale and the base state velocity $u_0$ as velocity scale. This introduces the Froude number $F = u_0^2/(gr)$, $g$ being the gravity constant, and the Reynolds number $R = \rho_p r u_0/\mu_p$ as dimensionless parameters; for abbreviation I set $\mu = F/R$. In addition, it is natural to introduce the density and viscosity ratios $\delta = \rho_c/\rho_p$ and $\nu = \mu_c/\mu_p$, as well as other viscosity-related coefficients $\kappa$ and $\bar{\kappa}$.

The absolute value of the base-state relative velocity between the two phases, $u_0$, is related to the density of the uniform base state $\phi_0$; within the above scaling it follows from $|1-\delta|\phi_0^{n+1} = 9\nu F/(2R)$. Thereby Stokesian drag and a Richardson- Zaki relation with exponent $n$ have been used [7]. As the model equations are Galilei-invariant, the base state may be chosen such that the particulate phase is at rest and the continuous phase is moving uniformly up- or downwards with velocity $u_0$ depending on whether a suspension of particles in a fluidized bed (with $\delta < 1$) or the flow of gas bubbles through a liquid column ($\delta > 1$) is considered. Thus, on the scale of $r$ and $u_0$ the base state is given by

$$(2.5)\quad \phi \equiv \phi_0,\ \nabla p_0 = -[1-\phi_0(1-\delta)]\mathbf{k},\ \mathbf{v}_0 = \mathbf{0},\ \mathbf{u}_0 = s\mathbf{k},\ s = \operatorname{sgn}(1-\delta).$$

To match this, the drag coefficient may be approximated by

$$(2.6)\qquad B(\phi) = |1-\delta|\frac{1-\phi}{\phi^n}\phi_0^{n+1}\,.$$

More complicated expressions for the drag, interparticle and viscous forces could be incorporated. For gas-liquid flows, the effects of virtual mass and lift forces should also be considered. This would make the analysis much more cumbersome and is therefore left to future work.

**3. Stability of the base state and a hierarchy of waves.** Thanks to certain simplifications inherent in the above equations, the linear stability analysis of the uniform state reduces to the investigation of the voidage perturbation equation

$$\begin{aligned} &\frac{A}{E}\left[\partial_t + \left(sc + \sqrt{f(c)}\right)\partial_x\right]\left[\partial_t + \left(sc - \sqrt{f(c)}\right)\partial_x\right]\phi \\ &+(\partial_t + s\,d\,\partial_x)\phi \;=\; E^{-1}\left[Am\partial_y^2 + J\left(\partial_t + sh\partial_x\right)\Delta\right]\phi, \end{aligned} \tag{3.1}$$

with the positive constants

$$\begin{aligned} &A = \phi_0 + \delta(1-\phi_0),\ c = \frac{\delta(1-\phi_0)}{A},\ E = \frac{|1-\delta|}{F},\ m = \frac{\phi_0|G(\phi_0)|}{A}, \\ &d = [B_0/\phi_0 - B_0' + |1-\delta|(1-2\phi_0)]/|1-\delta| = (n+2)(1-\phi_0), \\ &J = \frac{\phi_0(1+\kappa)}{R(1-\phi_0)} + H\,,\ \ H = \frac{\nu}{R}\frac{(1-\phi_0)(1+\bar{\kappa})}{\phi_0},\ \ h = \frac{H}{J}, \end{aligned} \tag{3.2}$$

and the function

$$f(\omega) = m - c(1-c) - (\omega - c)^2\,. \tag{3.3}$$

It should be noted that $E^{-1} \sim F$ which may be considered small in many practical circumstances and hence leads to a discussion of (3.1) in terms of wave hierarchies [8,9]. In the limit $F \to 0$, equation (3.1) reduces to a linearized kinematic wave equation, so that in first approximation disturbances of the uniform state travel with the kinematic wave velocity $sd$. For further discussion, let us first neglect the regularizing terms on the right hand side of (3.1). The equation lhs(3.1)=0 is hyperbolic if $f(c) > 0$, a condition that should be met in most practical cases. Then there are two dynamic waves travelling with speed $\pm\sqrt{f(c)}$ relative to the mean velocity $sc$; one is usually damped but in principle both interact with the kinematic wave. The base state is stable if the condition

$$sc - \sqrt{f(c)} < sd < sc + \sqrt{f(c)} \tag{3.4}$$

ensures that the kinematic wavefront lies within the domain of dependence cut by the characteristics of the dynamic waves. In this case the dynamic waves are both damped out rapidly so that the kinematic wave dominates the long-time behaviour.

If the base state is unstable, however, periodic travelling waves will dominate. In this case the kinematic wavefront approaches the persistent dynamic wave but cannot overtake it. This leads to fast growing high frequency waves and eventually, on the nonlinear level, to the breaking of the wave [10]. The transition is smoothed out and the disturbances evolve into a (stable or unstable) wavetrain if the terms on the right hand side of (3.1) are taken into account [11-13]. In addition, the inclusion of the transverse terms allows for the occurrence of oblique travelling waves and more complicated waves with nontrivial transverse structure. Small-amplitude waves near the stability limit are governed by perturbations of the Burgers-Korteweg deVries equation with either diffusion or anti-diffusion in the vertical direction [9,14-16]. In the amplifying case they reflect the competition between the long kinematic wave, which develops into a solitary wave, and a family of periodic travelling waves [16]. So far, however, these approximations have not been able to predict stable periodic or solitary wave solutions in fluidized beds [17].

The general stability condition following from (3.1) reads [6,18]

$$f(d) \geq 0, \qquad f(h) \geq 0, \tag{3.5}$$

meaning that the interparticle interaction must not be too weak compared to drag (measured relative to gravity) and "relative" viscous forces. The first condition corresponds exactly to (3.4), the second one arises from the inclusion of the rhs(3.1). The two conditions are not independent as can be seen from the relation

$$f(h) - f(d) = (d-h)(d+h-2c). \tag{3.6}$$

Often $f(d) \geq 0$ will be sufficient to ensure stability; in the dense region $\phi_0 \leq n/(n+2)$, for instance, $d > h$ and $d > 2c$ hold, so that $f(h) > f(d)$.

The origin of the conditions (3.5) is readily understood by considering the long- and short-wave limits. Equation (3.1) yields a quadratic dispersion relation for two complex eigenvalues $\sigma$ in dependence of longitudinal and transverse wavenumbers $\lambda$ and $k$, respectively. (By changing $\lambda$ into $-\lambda$ the complex-conjugate eigenvalues are obtained; replacing $k$ by $-k$ doesn't change anything because only $k^2$ enters the dispersion relation due to a reflectional symmetry of the equations, see below.) Here it is sufficient to restrict the discussion to one-dimensional perturbations, because the fastest growing disturbance is usually a plane wave moving into the vertical direction. In the long-wave limit, the real part of one eigenvalue starts out of zero at $\lambda = 0$ with $-Af(d)\lambda^2/E$, while it tends to the constant $-Af(h)/J$ as $\lambda \to \infty$. Therefore, an instability can occur only if at least one of the conditions (3.5) is violated. In this case, the most unstable wave is moving upwards if $s = 1$ and downwards if $s = -1$ (relative to the dispersed phase). The other eigenvalue has always negative real part and corresponds to a damped wave moving into the opposite direction except for large $\lambda$. Typical cases are depicted in Fig. 3.1. The kinematic wave occurs at $\lambda = 0$, up to two dynamic waves bifurcate at the points where $\mathrm{Re}(\sigma) = 0$ [19]. No small-amplitude bifurcation to pure one-dimensional waves takes place if $f(c) < 0$; in this case the base state is unstable to arbitrary perturbations. The possible presence of a short-wave instability should not be worried about too much as it might not occur in realistic situations [20]; it can of course be suppressed by adding regularizing higher-order terms to equation (2.1) or (2.3) [21,22].

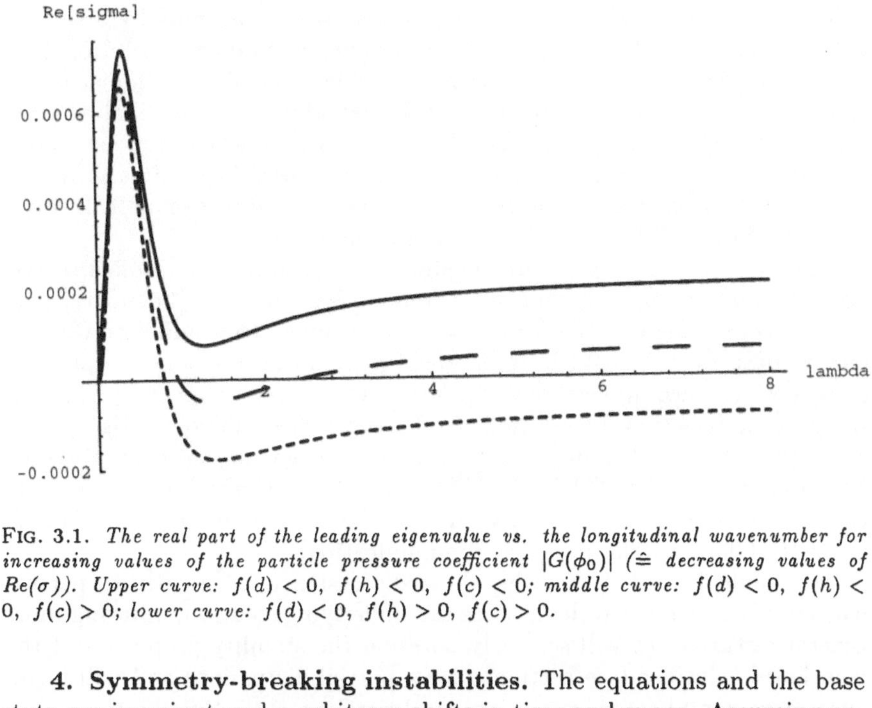


FIG. 3.1. *The real part of the leading eigenvalue vs. the longitudinal wavenumber for increasing values of the particle pressure coefficient* $|G(\phi_0)|$ *(*$\hat{=}$ *decreasing values of* $Re(\sigma)$*). Upper curve:* $f(d) < 0$, $f(h) < 0$, $f(c) < 0$*; middle curve:* $f(d) < 0$, $f(h) < 0$, $f(c) > 0$*; lower curve:* $f(d) < 0$, $f(h) > 0$, $f(c) > 0$.

**4. Symmetry-breaking instabilities.** The equations and the base state are invariant under arbitrary shifts in time and space. Assuming periodic boundary conditions and looking for time-periodic solutions, these translational symmetries transform into rotational symmetries. In addition, there exists a reflectional symmetry $\kappa_y : y \to -y$ in the horizontal direction, so that the symmetry group is given by $O(2)_y \times SO(2)_x \times S^1_t$. The oscillatory instability of the base state breaks the $SO(2)_x \times S^1_t$-symmetry down to a diagonal spatio-temporal symmetry, which is reflected in the bifurcation of periodic waves travelling into the vertical direction (with a possible sideways component if also the $SO(2)_y$ symmetry is broken). It is thus more convenient to perform the following discussion of successive symmetry breakings in the coordinate system $(z = x - s\omega t, y, t)$ of vertically travelling waves [6]. Then the first bifurcation is of stationary type and breaks only the spatial symmetries of the underlying group $O(2)_y \times SO(2)_z \times S^1_t$. Except for a small parameter regime the main instability occurs in the vertical direction; this breaks the $SO(2)_z$ symmetry. A further one-dimensional instability would break the time symmetry but leave the horizontal symmetries intact; such a solution corresponds to a quasi-periodic wave in the original coordinate system. Other instabilities along the uniform state lead directly to two-dimensional travelling waves. There are four of them: a pair of oblique travelling waves which obey a diagonal $SO(2)_{(y,z)}$ symmetry in space but break the reflectional symmetry in $y$ (they can be transformed into each other by a reflection at the $z$-axis), and a pair of "stationary" (w.r.t. $y$) travelling waves, one being symmetric and the other being anti-symmetric (because they differ only by a shift of half a period in $y$, they are usually counted as one branch of solutions). A similar set of solutions emerges from a secondary instability against transverse disturbances of the plane wave if this instability is again of stationary type. If it is oscillatory, time-periodic solutions arise which may be coined "rotating" and "standing" travelling waves because they represent waves which either travel simultaneously in the transverse direction or still travel vertically but with horizontally oscillating amplitude.

These waves still obey some symmetries which may be broken in tertiary bifurcations caused, e.g., by the interaction between different types of solutions emerging from the same bifurcation point. Depending on the secondary pattern, this will lead to simple time-periodic or more complicated quasi-periodic two-dimensional solutions. A summary is given in Fig. 4.1. In addition, it is conceivable that similar patterns of different origin interact with each other leading to more solutions. It has indeed been observed numerically that the STW and S'TW branches interact for small amplitudes if their bifurcation points are close to each other, but later break apart [B. Glasser 1995, personal communication].

It should be noted that the above discussion refers tacitly to perturbations of the same wavelengths as the underlying solution; allowing more general disturbances will seriously question the stability properties of the mentioned solutions (see section 6 for a partial discussion) and will introduce more complicated spatio-temporal patterns [17]. Finally, the condition of (finite) periodicity has also to be relaxed in order to include solitary waves [16,19]. The outlined picture can nevertheless serve as a backbone guide. It is valid for quite general assumptions on the closure relations but may differ to some extent when principal changes are made, e.g., by allowing an additional dependency of the particle pressure on the relative velocity of the two phases [23].

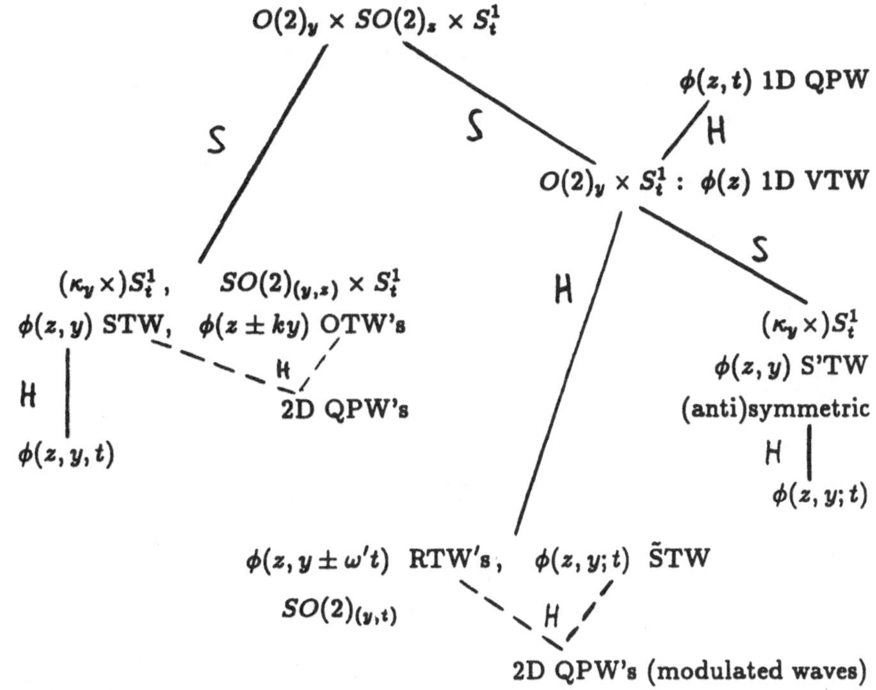


FIG. 4.1. *Outline of the most important symmetry-breaking bifurcations. S: stationary bifurcation, H: Hopf bifurcation.*

**5. Primary instabilities.** Looking for quasi-stationary solutions, i.e. solutions depending only on $z$ and $y$ but not on $t$, which bifurcate out of the base state, leads to relations between the longitudinal and transverse wavenumbers and the wave velocity $\omega$ which have to be satisfied at the bifurcation point [6]:

$$(5.1) \qquad \lambda^2 = \frac{Em(d-\omega)}{J(\omega-h)(m-f(\omega))}, \qquad k^2 = \frac{-E(d-\omega)f(\omega)}{J(\omega-h)(m-f(\omega))}.$$

These relations can only be satisfied if the right hand sides are positive. The first condition restricts $\omega$ to values between $h$ and $d$, then the second one decides for which parameter values bifurcations to one- or two-dimensional waves can occur. This is illustrated in Fig. 5.1 which is the equivalent to Fig. 3.1. Stabilization occurs at $\omega_{\max} = d$, while the singular behaviour at $\omega_{\min} = h$, which is close to zero in the case shown, allows for the onset of periodic solutions with arbitrarily large wavenumbers if $f(h) < 0$. It is evident that for small values of the particle pressure (for which $f(c) < 0$) no bifurcation to vertically travelling plane waves can take place but only to waves with a transverse structure. For intermediate values of $|G(\phi_0)|$ two one-dimensional waves bifurcate with different wave speeds $\omega_1 > \omega_2$ and wavenumbers $\lambda_1 < \lambda_2$, with no bifurcation to a two-dimensional wave between $\omega_1$ and $\omega_2$. For larger values of $|G(\phi_0)|$, $f(h)$ becomes positive and there is only one plane-wave bifurcation because $\omega_2$ has moved below $h$, while $\omega_1$ is increasing.

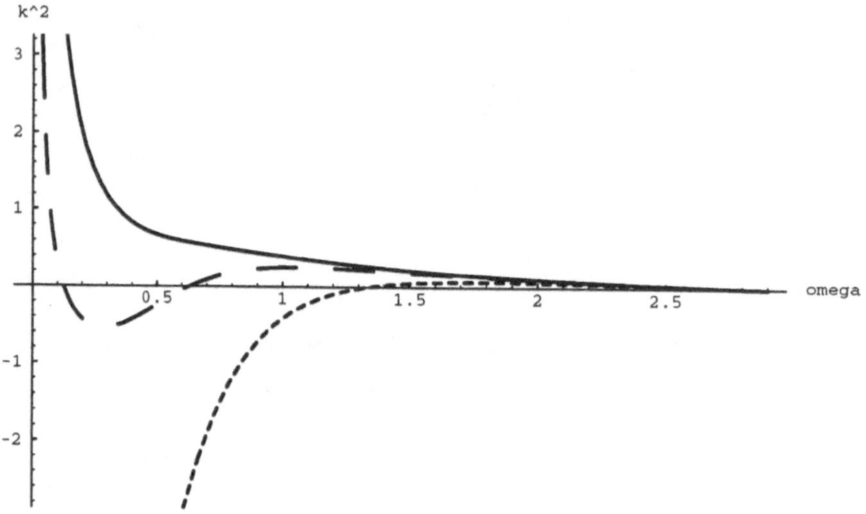


FIG. 5.1. *Evaluation of the second relation of (5.1) for different* $|G(\phi_0)|$ *but otherwise fixed parameters. The curves correspond to those of Fig. 3.1.*

Because $\lambda_1, k \to 0$ as $\omega_1 \to d$, one might expect the presence of another steady state or a solitary wave in that region. This is indeed the case, although no genuine two-dimensional solitary wave has been found so far. Vertically or obliquely travelling plane waves $\phi(x - s\omega t \pm \tilde{k}y)$, with $\tilde{k} = k/\lambda$, can be described by a second-order ordinary differential equation; thereby the mass conservation equations (2.1), (2.2) have been integrated once and the integration constants have been specified by imposing appropriate conditions on the volumetric fluxes of the two phases [24]. This allows for two more steady states which occur in a saddle-node bifurcation. One of these nontrivial steady states crosses the base state transcritically at $\omega = d$; its interaction with the periodic solution bifurcating at $\omega_1$ ($< d$ if $h < d$) produces a homoclinic solution, i.e. a solitary wave. As in the relations (5.1), the interplay between the viscous forces within each of the phases causes a singularity in the ODE, which is present for each $\omega \in [0, 1]$ and occurs at $\phi = \phi_0$ for $\omega = h$. Thus, the periodic solutions are confined to a region between the steady states and the singularity. They bifurcate from $\phi_0$ at

$\omega_{1,2} = c \pm \sqrt{m\tilde{k}^2 + f(c)}$ if the overall condition $m\tilde{k}^2 + f(c) > 0$ is satisfied as well as $m\tilde{k}^2 + f(d) < 0$ for a bifurcation at $\omega_1$, and $m\tilde{k}^2 + f(h) < 0$ for a bifurcation at $\omega_2$ [19].

In leading approximation, the periodic solution bifurcating at $\omega_1$ is given by $\phi = \phi_0 + \epsilon\phi_1 + O(\epsilon^2)$ (and similarly for the other variables), where $\epsilon$ denotes the amplitude of the wave and

$$(5.2) \qquad \phi_1 = 2\cos(z - \theta), \quad v_1^z = -\frac{s\omega_1}{1-\phi_0}\phi_1, \quad u_1^z = \frac{s(\omega_1 - 1)}{\phi_0}\phi_1,$$

with an arbitrary phase $\theta$; in addition $\lambda_1 \nabla p_1 \sim \phi_1 \mathbf{k}$. Therefore, in regions where the volume fraction of the fluid is perturbed to higher values, the fluid moves faster into the direction of the uniform flow if $\omega_1 > 1$ and slower if $\omega_1 < 1$, while the dispersed phase moves into the opposite direction. In particular, $\omega_1 < 1$ in dilute flows with $d \leq 1$, while $\omega_1 > 1$ in not too unstable dense fluidized beds.

**6. Secondary instability.** A lot of effort has been, and is being dedicated to the understanding of segregation phenomena in particulate systems. With regard to fluidized beds, there is often a striking difference between gas- and liquid-fluidized beds. While regions of high gas volume fraction form easily in gas-fluidized beds, liquid-fluidized beds seem less prone to bubbling. The observation of rising concentration waves in liquid-fluidized beds, which are at first planar but subsequently develop a moderate transverse structure [25,26], has led to the suggestion that a similar but stronger mechanism is at work in gas-fluidized beds that prevents the visible appearance of plane waves and leads directly to two- or three-dimensional voids (in wide/deep beds). A good candidate for the latter would be the two-dimensional STW wave that bifurcates directly from the uniform state; however, it usually comes out unstable and appears to be too symmetric in the vertical direction. Although this might change for higher amplitudes, there is still to cope with the presence of a possibly stable one-dimensional solution. So the emphasis has been laid mainly on a transverse instability of plane waves in both gas- and liquid-fluidized beds; nevertheless, the STW modes play a very active role in the generation of this secondary instability.

Besides the STW/OTW "mixed" modes there are two other modes helping to render the plane wave unstable. These start as stable pure transverse modes at the primary bifurcation point and remain stable along the uniform state for all parameter values, but gain longitudinal structure and may indeed become unstable as the plane wave grows in amplitude. Simultaneously the mixed modes obtain more transverse structure, so that eventually all these "least-stable" perturbation modes are of similar type, namely bubble-like, and act on the same scale. The interaction between this disturbance packet and the plane wave finally triggers the onset of the secondary instability [20,27].

In an infinite bed, the secondary instability would set in right away; but since we always have to do with finite configurations, the plane wavetrain has a chance to grow to a small but finite amplitude before this happens. The critical amplitude depends on the width of the container, because this restricts the transverse wavelength of admissible two-dimensional perturbations. For small amplitudes of the plane wave, the eigenmodes of the critical eigenvalue at the onset of an S'TW look very much like those of the STW [6], except for an additional horizontal variation of the vertical velocities [27]. This last contribution is crucial as it leads to alternating columns of up- and downwards moving phases, with a change of direction every half wavelength horizontally, and a form of tilting and sliding mechanism similar to the one described for stratified fluids [28,29]. The voidage perturbation is not only cellular but also phase-shifted with respect to the primary wave, thus the previously planar wavefronts are bent and the two phases are slightly pushed sidewards. This leads to a transverse segregation of fluid and particles and, at a later stage, to the formation of a "hole" [30]; whether this hole develops into a void or a cluster depends on the voidage value of the base state.

From a bifurcation point of view the behaviour of gas- and liquid-fluidized beds is very similar, but great quantitative differences exist [20,31]. If the bed is only weakly unstable, the primary bifurcation is subcritical but the branch of one-dimensional solutions turns around and gains stability for higher amplitudes; for more unstable beds the bifurcation is supercritical and the periodic plane waves seem to stay one-dimensionally stable up to large amplitude. In any case a secondary instability to two-dimensional travelling waves sets in once the amplitude of the plane wave has become large enough; this bifurcation may be sub- or supercritical but eventually leads to a stable (with respect to perturbations with the same symmetry and wavelength) solution [30,31]. Direct numerical simulations, however, show that initial perturbations of the uniform state do not necessarily develop into the two-dimensional travelling wave for systems which are known as non-bubbling [31,32]. This suggests that in such a case another nearby solution exists to which the S'TW loses its stability. A "semi-theoretical" criterion to distinguish bubbling from non-bubbling beds has been developed in [20,31] based on the relative magnitude of primary and secondary growth rates, which may be traced back to the old empirical Froude number criterion of Wilhelm & Kwauk [33].

**Acknowledgements.** I am grateful to B. J. Glasser, R. Jackson, Y. G. Kevrekidis and S. Sundaresan for many fruitful conversations, and the Deutsche Forschungsgemeinschaft for continuing financial support.

## REFERENCES


[1] G. Hetsroni (ed.), Handbook of multiphase systems (McGraw-Hill, New York, 1982).

[2] D.A. Drew, *Mathematical modelling of two-phase flow*, Ann. Rev. Fluid Mech. 15 (1983), pp. 261–291.

[3] M. Golubitsky, I. Stewart, and D.G. Schaeffer, Singularities and groups in bifurcation theory, Vol. II, AMS 69 (Springer, Berlin, 1988).

[4] R. Jackson, *Fluid mechanical theory*, in: J.F. Davidson and D. Harrison (eds.), Fluidization (Academic Press, London, 1971), pp. 65–119.

[5] S.K. Garg and J.W. Pritchett, *Dynamics of gas- fluidized beds*, J. Appl. Phys. 46 (1975), 4493–4500.

[6] M.F. Göz, *On the origin of wave patterns in fluidized beds*, J. Fluid Mech. 240 (1992), 370–404.

[7] J.F. Richardson, *Incipient fluidization and particulate systems*, in: J.F. Davidson and D. Harrison (eds.), Fluidization (Academic Press, London, 1971), pp. 26–64.

[8] J.T.C. Liu, *Note on a wave-hierarchy interpretation of fluidized bed instabilities*, Proc. Roy. Soc. A 380 (1982), 229–239.

[9] A. Kluwick, *Small-amplitude finite-rate waves in suspensions of particles in fluids*, Z. Angew. Math. Mech. 63 (1983), 161–171.

[10] G.H. Ganser and J.H. Lightbourne, *Oscillatory traveling waves in a hyperbolic model of a fluidized bed*, Chem. Engng. Sci. 46 (1991), 1339–1347.

[11] J.T.C. Liu, *Nonlinear unstable wave disturbances in fluidized beds*, Proc. Roy. Soc. A 389 (1983), 331–347.

[12] G.H. Ganser and D.A. Drew, *Nonlinear periodic waves in a two-phase flow model*, SIAM J. Appl. Math. 47 (1987), 726–736.

[13] G.H. Ganser and D.A. Drew, *Nonlinear stability analysis of a uniformly fluidized bed*, Intl. J. Multiphase Flow 16 (1990), 447–460.

[14] T.S. Komatsu and H. Hayakawa, *Nonlinear waves in fluidized beds*, Phys. Lett. A 183 (1993), 56–62.

[15] S.E. Harris and D.G. Crighton, *Solitons, solitary waves and voidage disturbances in gas-fluidized beds*, J. Fluid Mech. 266 (1994), 243–276.

[16] M.F. Göz, *Small Froude number asymptotics in two- dimensional two-phase flows*, Phys. Rev. E 52 (1995), 3697–3710.

[17] M.F. Göz, *A "complete" two-dimensional nonlinear wave equation describing the small Froude number regime of multiphase flows*, in: L. Pust and F. Peterka (eds.), Proceedings of the 2nd European Nonlinear Oscillations Conference (Prague 1996), Vol. 1, 183–186.

[18] M.F. Göz, B.J. Glasser, Y.G. Kevrekidis and S. Sundaresan, *Traveling waves in multi-phase flows*, in: M. Rahman and C.A. Brebbia (eds.), Advances in Fluid Mechanics, Comp. Mech. Publ. (South Hampton/Boston 1996), 307–316.

[19] M.F. Göz, *Bifurcation of plane voidage waves in fluidized beds*, Physica D 65 (1993), 319–351.

[20] M.F. Göz and S. Sundaresan, *The growth and saturation of one- and two-dimensional disturbances in fluidized beds*, submitted to J. Fluid Mech.

[21] H. Hayakawa, T.S. Komatsu, and T. Tsuzuki, *Pseudo-solitons in fluidized beds*, Physica A 204 (1994), 277–289.

[22] M.F. Göz, *Quasi-stationary instabilities in fluidized beds*, Phys. Lett. A 200 (1995), 355–359.

[23] J.A. Hernandez and J. Jimenez, *Bubble formation in dense fluidised beds*, in: J. Jimenez (ed.), Proc. NATO Advanced Research Workshop on the Global Geometry of Turbulence (Plenum 1991), pp. 133–142.

[24] M. F. Göz, *Instabilities and the formation of wave patterns in fluidized beds*, in: G. Gouesbet and A. Berlemont (eds.), Instabilities in Multiphase Flows (Plenum 1993), pp. 251–259.

[25] A.K. Didwania and G.M. Homsy, *Flow regimes and flow transitions in liquid fluidized beds*, Intl. J. Multiphase Flow 7 (1981), 563–580.

[26] A.K. Didwania and G.M. Homsy, *Resonant side-band instabilities in wave propagation in fluidized beds*, J. Fluid Mech. 122 (1982), 433–438.

[27] M.F. Göz, *Transverse instability of plane wavetrains in gas-fluidized beds*, J. Fluid Mech. 303 (1995), 55–82.

[28] G.K. Batchelor and J.M. Nitsche, *Instability of stationary unbounded stratified fluid*, J. Fluid Mech. 227 (1991), 357–391.

[29] G.K. Batchelor, *Secondary instability of a gas fluidized bed*, J. Fluid Mech. 257 (1993), 359–371.

[30] B.J. Glasser, I.G. Kevrekidis, and S. Sundaresan, *One- and two-dimensional travelling wave solutions in gas-fluidized beds*, J. Fluid Mech. 306 (1996), 183–221.

[31] B.J. Glasser, I.G. Kevrekidis, and S. Sundaresan, *Fully-developed traveling wave solutions and bubble formation in fluidized beds*, to appear in J. Fluid Mech.

[32] K. Anderson, S. Sundaresan, and R. Jackson, *Instabilities and the formation of bubbles in fluidized beds*, J. Fluid Mech. 303 (1995), 327–366.

[33] R.H. Wilhelm and M. Kwauk, *Fluidization of solid particles*, Chem. Engng. Progr. 44 (1948), 201–218.

# THREE DIMENSIONAL VISCOELASTICITY IN FINITE STRAIN: FORMULATION OF A RATE-TYPE CONSTITUTIVE LAW CONSISTENT WITH DISSIPATION

MANSOOR A. HAIDER* AND MARK H. HOLMES*

**Abstract.** Viscoelastic constitutive models can predict transient phenomena like creep and stress relaxation. Among the materials that can exhibit such effects in finite strain are biological soft tissues which are commonly modeled using a multiphasic continuum theory. Under infinitesimal strain, the classical 1-D Standard Linear Model (1-D SLM) is a simple law containing a stress rate and exhibiting the desired transient and equilibrium behavior observable in many soft tissues. The derivation of a rate-type constitutive law appropriate for modeling the non-linear viscoelasticity of soft tissues is the focus of this study. Well-posed laws should be objective and consistent with thermodynamic considerations of dissipation and energy. Infinitesimal models are not objective, while many non-linear analogies to the 1-D SLM fail to address dissipation. In the current study, internal variables are introduced, and employed in the derivation of a 3-D non-linear rate-type viscoelastic constitutive law. Evolution of the internal variables is assumed to involve first order rates. Properties of the 1-D SLM as well as existing non-linear models of soft tissues are used to motivate the constitutive assumptions and additional requirements. These requirements include symmetry of the stress, isotropy, reduction to hyperelasticity (via material parameters) and the existence of a hyperelastic equilibrium state. A class of objective rate-type constitutive laws satisfying dissipation and the additional requirements is derived. As an illustration, a compressible finite linear model is formulated. In infinitesimal strain, this model provides a 3-D analogy to the 1-D SLM with a set of constraints on the material parameters. The finite linear model is analyzed under simple time-dependent compression, extension and shear and shown to be consistent with expected behavior.

**Key words.** non-linear viscoelasticity, hyperelasticity, dissipation, soft tissue.

**1. Introduction.** Viscoelastic constitutive models can predict transient phenomena like *creep* and *stress relaxation* [6]. Single phase materials, such as rubber polymers, as well as multiphasic materials, like biological soft tissues, can exhibit such effects in finite strain. Generally speaking, viscoelastic laws can be separated into the (non-exclusive) categories of integral and rate-type models. For instance, in infinitesimal strain, one dimensional integral laws can be written in the form:

$$\sigma(z,t) = -\int_0^\infty G(s)\frac{\partial}{\partial s}\left(\frac{\partial u}{\partial z}(z,t-s)\right)ds \tag{1.1}$$

A typical rate-type law is the *standard linear model* which can be written in the form:

$$\sigma(z,t) + \tau_0^\sigma \frac{\partial \sigma}{\partial t}(z,t) = H_A\left(\frac{\partial u}{\partial z}(z,t) + \tau_0^u \frac{\partial^2 u}{\partial t \partial z}(z,t)\right) \tag{1.2}$$

* Dept. of Mathematics, Duke University, Box 90320, Durham, NC 27708.
This work has been supported by the National Science Foundation under a Graduate Research Fellowship and grants ASC-9318184 and DMS-9404517.

where $H_A$, $\tau_0^\sigma$ and $\tau_0^u$ are material constants.

Methods for formulating non-linear viscoelastic constitutive laws range from *a priori* generalizations of (1.1) or (1.2), to the construction of macroscopic laws from a detailed analysis of microstructure. Regardless of approach, the resulting laws should satisfy fundamental postulates of nonlinear constitutive modeling. Two such postulates are *material frame indifference* (or *objectivity*) and consistency with the *Second Law of Thermodynamics* (or *dissipation*). Many objective laws have been formulated by generalizing (1.1) or (1.2) to the case of three dimensional finite strain using appropriate stress, strain and rate tensors from the theory of nonlinear elasticity [15,16]. However, most of these formulations fail to address the issue of consistency with dissipation. A few exceptions are now briefly described.

In the case of infinitesimal strain, conditions on the relaxation function $G(s)$ in (1.1) that are necessary and sufficient for dissipation, in the form of the *Clausius-Duhem inequality*, have been presented in [5]. However, we are unaware of any extension of these conditions to non-linear integral models such as the *Theory of Finite Linear Viscoelasticity* [4]. In [19], the following class of constitutive laws, written in the material frame, are considered:

$$\dot{\mathbf{S}} - \mathcal{X}(\mathbf{E}, \mathbf{S})\dot{\mathbf{E}} = \mathbf{H}(\mathbf{E}, \mathbf{S}) \tag{1.3}$$

Here $\mathbf{S}$ is the *second Piola-Kirchhoff stress tensor*, $\mathbf{E}$ is the *Lagrangian strain tensor*, and $\mathbf{H}$ and $\mathcal{X}$ are (respectively) second and fourth-rank tensor functions. Specifically, the circumstances under which the rate-type law (1.3) is compatible with a free energy function satisfying dissipation are considered in [19]. However, in the course of the analysis, the fourth-rank tensor $\mathcal{X}$ is restricted to be constant and, as such, a result is obtained only in this special case.

In the present study, we will formulate a rate-type constitutive law that is a sub-class of the general form (1.3) and satisfies both objectivity and dissipation. Our motivation is the extension of a biphasic continuum model of soft tissues to incorporate viscoelasticity into the solid tissue matrix. We choose to formulate a rate-type law in the material frame, as such an approach has many computational advantages in the numerical simulations of complex three dimensional joint mechanics. For this initial extension of the biphasic model, it is appropriate to require that the constitutive laws be symmetric, isotropic, hyperelastic in steady state, and reducible to hyperelasticity via a choice of the material constants. In §2, the biphasic model is briefly described and it is demonstrated that a formulation accounting for intrinsic viscoelasticity reduces to a consideration of viscoelasticity for a single solid phase. The simplifying assumptions and requirements to be imposed on the constitutive functions are described in §3. The approach used is similar to that of [3] in that it involves the derivation of constitutive relations from free energy potentials using internal variables. Motivated by the standard linear model (1.2), we will assume that the stress and free energy potentials can be decomposed additively into stored and dissipative parts [12,13,14,22]. Although our approach requires that ***a priori*** choices be made, these choices are made for the free energy potentials and thus allow for the formulation of a rate-type law satisfying dissipation and certain energy considerations. In §4, we present a ***finite linear model*** which yields a rate-type law that is a sub-class of the general form (1.3), without the restriction in [19] that $\mathcal{X}$ need be constant. The resulting law involves three viscoelastic material constants and is shown, in §5, to yield a stress response that is consistent with expected behavior under several simple transient deformations.

## 2. Biphasic model of soft tissues.

**2.1. Background.** *Articular cartilage* is a white connective tissue covering the articulating surfaces of joints in the knee, hip and shoulder. A distinguishing characteristic of cartilage is that a large proportion of its content is unbound water. Its principal structural component is a porous *extra-cellular matrix (ECM)* consisting of *collagen* fibers that are intertwined with enormous macromolecules called *proteoglycans*. The more recent mathematical models of soft tissues like cartilage are based on the use of a *biphasic continuum model* [8,17]. The biphasic model is similar to classical poroelastic models [1,23] and is rooted in the modern mixture theories [2,25]. In the biphasic model, inertia is neglected and the tissue is idealized to be a two-phase incompressible mixture consisting of a solid ECM saturated by a fluid phase. In the non-linear model [8,10], the solid phase is taken to be hyperelastic, the fluid phase inviscid, and there is a fluid-solid drag with a permeability that is strain-dependent. Such a model accounts for creep and stress relaxation at the macroscopic level via *flow-dependent viscoelasticity* due to the fluid-solid drag.

Our motivation is to extend this model to account for non-linear viscoelasticity in the ECM. In the context of the biphasic model, this is commonly referred to as *matrix viscoelasticity* or *intrinsic viscoelasticity* as it is confined to the solid phase. Evidence of an intrinsic viscoelastic effect arises from experiments in torsional shear [7,20]. In the case of infinitesimal strain, initial studies have incorporated intrinsic viscoelasticity into the biphasic model and demonstrated improved accuracy and consistency in comparison to experimental data [9,21]. This study is motivated by a need to extend these ideas to the case of finite strain.

**2.2. Balance laws and the entropy inequality.** For a biphasic mixture in finite strain, there are kinematic relations that account for the motion of each of the two phases. A *reference configuration* is defined for each phase by $\Omega_0^{(s)}, \Omega_0^{(f)} \subset \mathcal{R}^3$ with *material coordinates* (respectively) $\mathbf{X}^{(s)} \in \Omega_0^{(s)}, \mathbf{X}^{(f)} \in \Omega_0^{(f)}$. *Deformations* $\chi^{(s)} : \Omega_0^{(s)} \times \mathcal{R} \to \mathcal{R}^3$ and $\chi^{(f)}$ :

$\Omega_0^{(f)} \times \mathcal{R} \to \mathcal{R}^3$ can then be defined for each phase as the smooth invertible mappings:

$$\text{(2.1)} \qquad \mathbf{x} = \mathbf{x}^{(s)} = \chi^{(s)}(\mathbf{X}^{(s)}, t) \qquad \mathbf{x} = \mathbf{x}^{(f)} = \chi^{(f)}(\mathbf{X}^{(f)}, t)$$

where the points $\mathbf{x}$ are the *spatial coordinates*. We note that (2.1) provides a description in which each spatial point $\mathbf{x}$ has, associated with it, a deformation for each of the two phases.

Associated with each phase are the *volume fractions* $\phi^{(s)}$ and $\phi^{(f)}$ which represent the ratio of phase to mixture volumes. Related to the volume fractions are the *phase densities* $\rho^{(s)}$ and $\rho^{(f)}$, measured per unit volume of the mixture, and the *true phase densities* $\rho_T^{(s)}$ and $\rho_T^{(f)}$, intrinsic to each phase. The densities and volume fractions are related by $\rho^{(s)} = \phi^{(s)}\rho_T^{(s)}$ and $\rho^{(f)} = \phi^{(f)}\rho_T^{(f)}$

These relations are supplemented by requirements that the mixture be *saturated* $\phi^{(s)} + \phi^{(f)} = 1$ and *intrinsically incompressible*, $\dot{\rho}_T^{(s)} = 0$ and $\dot{\rho}_T^{(f)} = 0$. The mixture is also assumed to be initially *homogeneous* so that the true phase densities are constant and incompressibility can be expressed via the single constraint:

$$\text{(2.2)} \qquad \nabla_x \cdot \left(\phi^{(s)}\mathbf{v}^{(s)} + \phi^{(f)}\mathbf{v}^{(f)}\right) = 0$$

Here $\mathbf{v}^{(s)}$ and $\mathbf{v}^{(f)}$ are the *phase velocities* and can be determined by time differentiation of (2.1).

Using (2.2), *balance of mass* for each phase can then be written as:

$$\text{(2.3)} \qquad \phi_0^{(s)} = J^{(s)}\phi^{(s)} \qquad \phi_0^{(f)} = J^{(f)}\phi^{(f)}$$

where $\phi_0^{(s)}, \phi_0^{(f)}$ are the (constant) volume fractions in the reference configuration and $J^{(s)}$, $J^{(f)}$ are the *Jacobians* associated with the deformations (2.1). *Balance of momentum* for each phase is written as:

$$\text{(2.4)} \qquad \nabla_x \cdot \sigma^{(s)} + \pi^{(s)} = \mathbf{0} \qquad \nabla_x \cdot \sigma^{(f)} + \pi^{(f)} = \mathbf{0}$$

where inertial terms have been neglected. In (2.4), $\sigma^{(s)}, \sigma^{(f)}$ are the *partial Cauchy stress tensors* for each phase and $\pi^{(s)}, \pi^{(f)}$ are *momentum exchange vectors* accounting for an interaction mechanism between the two phases. It is assumed that there is no moment-of-momentum supply so that balance of angular momentum gives symmetric partial stresses. Momentum balance for the mixture, as a whole, necessitates that $\pi^{(s)} = -\pi^{(f)}$.

If the solid deformations $\mathbf{x}^{(s)}$, fluid velocities $\mathbf{v}^{(f)}$ and volume fractions $\phi^{(s)}$, $\phi^{(f)}$ are taken as the set of dependent variables, then equations (2.3), (2.4) and $\phi^{(s)} + \phi^{(f)} = 1$ form a system of 9 equations in 8 unknowns. To form a closed system, the incompressibility condition (2.2) is introduced as a constraint with *Lagrange multiplier* $p$. It then remains to determine constitutive laws for the partial stresses $\sigma^{(s)}$, $\sigma^{(f)}$, and momentum exchange vector $\pi^{(s)}$. These constitutive laws can be constrained by requiring that they be consistent with the Second Law of Thermodynamics as expressed by a local form of the *Clausius-Duhem inequality.* For most soft tissues, it is appropriate to assume that a constant temperature is retained and that heat conduction is negligible. Under these assumptions, the entropy inequality for a biphasic mixture can be expressed in the following form [8]:

$$
\begin{aligned}
(2.5)\qquad & -\rho^{(s)}\dot{\Psi}^{(s)} - \rho^{(f)}\dot{\Psi}^{(f)} + tr\left[(\sigma^{(s)} + \phi^{(s)}p\mathbf{i})\mathbf{d}^{(s)} + (\sigma^{(f)} + \phi^{(f)}p\mathbf{i})\mathbf{d}^{(f)}\right] \\
& \qquad + (\mathbf{v}^{(s)} - \mathbf{v}^{(f)}) \cdot (p\nabla_x\phi^{(s)} - \pi^{(s)}) \geq 0
\end{aligned}
$$

Here, $\Psi^{(s)}$, $\Psi^{(f)}$ are the *free energies* and the *rate-of-deformation* tensors associated with each phase are denoted by $\mathbf{d}^{(s)}$,$\mathbf{d}^{(f)}$ (Note that $\mathbf{i}$ denotes the identity matrix in the Eulerian or spatial configuration). In obtaining (2.5), the condition of intrinsic incompressibility (2.2) has been introduced as a constraint with Lagrange multiplier $p$ and balance laws for momentum and energy have also been employed. The inequality (2.5) must then hold for all admissible processes and can be used to constrain the form of the constitutive laws.

**2.3. The reduced entropy inequality.** In formulating a set of constitutive laws for the biphasic model, we apply the *Principle of Phase Separation* [18] in place of the *Principle of Equipresence* [24]. "Separated" laws for each phase are then augmented by additional relations which account for the interaction mechanisms between the two phases of the mixture. The fluid phase is assumed to be inviscid and incompressible, and it can be shown that a constitutive law consistent with (2.5) is given by:

$$
(2.6)\qquad \sigma^{(f)} = -\phi^{(f)}p\mathbf{i} = -\frac{\rho^{(f)}}{\rho_T^{(f)}}p\mathbf{i}
$$

Similarly, a fluid-solid drag law satisfying the dissipation inequality (2.5) is given by:

$$
(2.7)\qquad \pi^{(s)} = p\nabla_x\phi^{(s)} - K(\mathbf{v}^{(s)} - \mathbf{v}^{(f)}) \qquad \text{where:} \quad K \geq 0
$$

$K$ is a scalar *diffusive drag coefficient* and, in the more general biphasic models, is assumed to be a non-linear function of the strain [10,11] .

It is noted that (2.7) accounts for a flow-dependent viscoelastic effect, and it remains to determine a constitutive law for the solid stress that accounts for intrinsic viscoelasticity. Since (2.5) must hold true when the drag coefficient $K$ is zero, we can substitute (2.6)-(2.7) into (2.5) and use a material description to obtain:

$$
(2.8)\qquad -\rho_0^{(s)}\dot{\Psi}^{(s)} + tr\left[\left(\frac{1}{2}\mathbf{S}^{(s)} + \frac{1}{2}pJ^{(s)}\phi^{(s)}(\mathbf{C}^{(s)})^{-1}\right)\dot{\mathbf{C}}^{(s)}\right] \geq 0
$$

Here, $\mathbf{S}^{(s)}$ is the second Piola-Kirchhoff stress tensor and $\mathbf{C}^{(s)}$ is the *right Cauchy-Green deformation* tensor [15,16]. Henceforth, they will be referred to as the material stress and material strain (respectively), and it is noted that $\mathbf{C}^{(s)}$ is related to the Lagrangian strain tensor $\mathbf{E}^{(s)}$ (see (1.3)) by $\mathbf{E}^{(s)} = 1/2(\mathbf{C}^{(s)} - \mathbf{I})$. We now write the constitutive function for the material stress of the solid phase as:

$$\mathbf{S}^{(s)} = -pJ^{(s)}\phi^{(s)}(\mathbf{C}^{(s)})^{-1} + \mathbf{S}_E \tag{2.9}$$

where $\mathbf{S}_E$ is called the *extra stress in the solid matrix*. Substitution of (2.9) into (2.8) reduces the inequality to:

$$-\rho_0^{(s)}\dot{\Psi}^{(s)} + tr\left[\frac{1}{2}\mathbf{S}_E\dot{\mathbf{C}}^{(s)}\right] \geq 0 \tag{2.10}$$

We then note that (2.10) is equivalent to the Clausius-Duhem (or dissipation) inequality for a single phase viscoelastic solid. As such, intrinsic viscoelasticity for the biphasic model can be addressed by formulating a constitutive law for a single phase viscoelastic solid that is consistent with (2.10). For convenience, we will now drop all superscripts associated with the solid phase variables and all subscripts associated with the extra stress.

## 3. Constitutive assumptions.

**3.1. Simplifying assumptions.** In the previous section, we have reduced our problem to one of determining constitutive functions for the free energy and stress of a viscoelastic solid that satisfy (2.10). If we restrict our solid to the class of *simple* and *homogeneous* materials [24] then it can be shown that, for satisfaction of objectivity, the stress-strain law assumes the general form [15]:

$$\mathbf{S} = \mathcal{M}\left[\mathbf{C}(\mathbf{X}, t)\right] \tag{3.1}$$

One way of achieving the functional dependence in (3.1) is to introduce internal variables and an additional evolution function [3]. Specifically, we introduce an *internal material tensor* $\mathbf{A}(\mathbf{X}, t)$ into the constitutive formulation and assume that it is governed by an *evolution function* $\dot{\mathbf{A}} = \bar{\mathbf{K}}$. It is noted that such an approach facilitates the formulation of a rate-type law in the class of (1.3).

The dependence of the constitutive functions on the kinematic and internal variables is motivated by the one dimensional infinitesimal standard linear model (1.2). The rate-type law (1.2) can be derived from a consideration of the physical element shown in Figure 1. Specifically, when the strain of the spring in the lower branch is introduced as an internal variable, then the stress and free energy can be additively decomposed into stored and dissipative parts. In addition, the force balance between the spring and dashpot in the lower branch provides an evolution function for

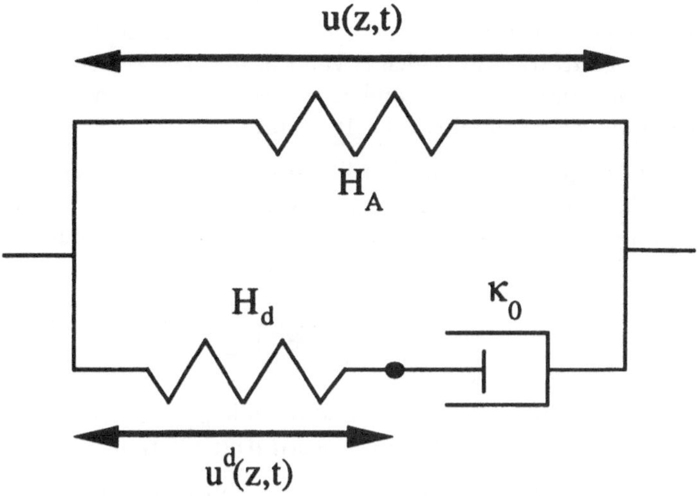


FIG. 1. *One dimensional standard linear model. The strain of the spring in the lower branch is introduced as an internal variable.*

the internal variable. Although there is no physical analogy to this element in either three dimensions or finite strain, the constitutive forms can be used to motivate the development of a finite strain law. Such a law will then be consistent with the standard linear model under the assumptions of infinitesimal strain and uniaxial motion.

Motivated by this, we choose the following forms for the stress, free energy and evolution functions:

$$\Psi = \bar{\Psi}(\mathbf{C}, \mathbf{A}) = \bar{\Psi}^s(\mathbf{C}) + \bar{\Psi}^d(\mathbf{A}) \qquad \mathbf{S} = \bar{\mathbf{S}}(\mathbf{C}, \mathbf{A}) = \bar{\mathbf{S}}^s(\mathbf{C}) + \bar{\mathbf{S}}^d(\mathbf{A}) \tag{3.2}$$
$$\dot{\mathbf{A}} = \bar{\mathbf{K}}(\dot{\mathbf{C}}, \mathbf{A})$$

Here, $\bar{\mathbf{S}}^d$, $\bar{\Psi}^d$ are called the *dissipative stress* and *dissipative energy functions* while $\bar{\mathbf{S}}^s$, $\bar{\Psi}^s$ are the called the *stored stress* and *stored energy functions* (respectively). The stored functions represent recoverable energy in the system that is present at equilibrium, while the dissipative functions represent energy that is lost, in steady state ($t \to \infty$), to internal dissipation. It is apparent that the assumptions in (3.2) restrict our analysis to a sub-class of the non-linear viscoelastic materials. It should be emphasized that, by no means, are these assumptions necessary for consistency with the standard linear model. However, as will be demonstrated in §4, these simplifications are useful in that they facilitate the development of a rate-type law that is of the form (1.3). We note that an additive decomposition has been used by several authors ([12,13,14,22]) to formulate viscoelastic laws in the case of finite strain.

**3.2. Additional requirements.** To facilitate a choice of the constitutive functions in (3.2), we now introduce some additional requirements. These requirements will allow us to derive a three dimensional rate-type viscoelastic constitutive law which is valid in finite strain, satisfies certain thermodynamic considerations, and is consistent with some fundamental properties observable in biological soft tissues.

**3.2.1. Second law of thermodynamics (dissipation).** We will require that the constitutive functions in (3.2) satisfy the Second Law of Thermodynamics (or dissipation) as represented by the reduced entropy inequality (2.10) which, in the new notation, is:

$$-\rho_0 \dot{\bar{\Psi}} + tr\left(\frac{1}{2}\bar{\mathbf{S}}\dot{\mathbf{C}}\right) \geq 0 \tag{3.3}$$

The inequality (3.3) must hold for all admissible processes at all times and will be used, in §4, to derive a set of constitutive relations.

**3.2.2. Considerations of energy.** We will require that the free energy function $\bar{\Psi}(\mathbf{C}, \mathbf{A})$ in (3.2) have several essential properties. We first define an *equilibrium state* as one in which $\dot{\mathbf{C}} \to \mathbf{0}$ as $t \to \infty$. The value of the material strain tensor $\mathbf{C}$ in such a state is independent of time and will be denoted by $\mathbf{C}_\infty \equiv \lim_{t\to\infty} \mathbf{C}$. We will first require that the components of the internal material tensor $\mathbf{A}$ assume the constant values $\mathbf{A}^{eq}$ in any equilibrium state, so that:

$$\mathbf{A} \to \mathbf{A}^{eq} \qquad \text{whenever:} \qquad \dot{\mathbf{C}} \to \mathbf{0} \qquad \text{as} \quad t \to \infty \tag{3.4}$$

Another requirement is that the free energy remain non-negative so that:

$$\bar{\Psi}(\mathbf{C}, \mathbf{A}) \geq 0 \qquad \text{for all admissible processes} \tag{3.5}$$

Ideally, we would like the free energy function to be "convex" in the neighborhood of a minimum energy state. However, imposing a convexity condition for a non-linear viscoelastic solid is beyond the scope of this study. The weaker condition (3.5) ensures that the free energy function is bounded below. Since, by (3.3), arbitrary constants may be added to the free energy without affecting the constitutive law for the stress, it suffices to take the minimum energy state as the zero state.

We will also require that all equilibrium states are *hyperelastic* [24] so that:

$$\bar{\Psi}(\mathbf{C}_\infty, \mathbf{A}^{eq}) = \bar{\Psi}^s(\mathbf{C}_\infty) \qquad \bar{\mathbf{S}}(\mathbf{C}_\infty, \mathbf{A}^{eq}) = \bar{\mathbf{S}}^s(\mathbf{C}_\infty) = 2\rho_0 \frac{\partial \bar{\Psi}^s}{\partial \mathbf{C}_\infty} \tag{3.6}$$

In (3.6), the $i, j$ component of $\partial\bar{\Psi}^s/\partial\mathbf{C}_\infty$ is given by $\partial\bar{\Psi}^s/\partial(\mathbf{C}_\infty)_{ij}$.

The set of requirements described by (3.3)-(3.6) can be loosely referred to as thermodynamic considerations. Their satisfaction in the derivation of our constitutive laws will prevent certain behaviors which are known to be physically implausible. We now consider a few additional requirements which can be regarded as simplifying assumptions appropriate for soft tissues.

**3.2.3. Reduction to hyperelasticity.** In modeling the viscoelastic behavior of soft tissues, the ability to determine values of the material constants in any constitutive law is of key concern. Given that the existing non-linear biphasic model [8] exhibits a hyperelastic equilibrium response, we distinguish between two types of material constants. *Elastic material constants* are those that can be determined from the equilibrium response of the tissue, while *viscoelastic material constants* require a consideration of the transient response. Of primary importance is the ability to quantify the viscoelastic effect via material constants and assess its relative contribution in comparison to other physical mechanisms. To facilitate this process, we will require that there exist a set of values of the viscoelastic material constants that reduce the model to hyperelasticity. These viscoelastic material constants will be denoted by $\gamma \equiv (\gamma_i), i = 1, ..., N_1$ and $\alpha \equiv (\alpha_j), j = 1, ..., N_2$. They will be included in only the dissipative constitutive functions of (3.2), which can then be written as $\bar{\mathbf{S}}^d(\mathbf{A};\gamma)$ and $\bar{\Psi}^d(\mathbf{A};\alpha)$. The requirement that there exist values $\gamma = \gamma^0$, $\delta = \delta^0$ such that the material is hyperelastic then becomes:

$$(3.7) \qquad \bar{\Psi}^d(\mathbf{A};\alpha^0) = 0 \qquad \bar{\mathbf{S}}^d(\mathbf{A};\gamma^0) = \mathbf{0} \qquad \bar{\mathbf{S}}^s(\mathbf{C}) = 2\rho_0 \frac{\partial \bar{\Psi}^s}{\partial \mathbf{C}}$$

**3.2.4. Consistency with the standard linear model.** In the preliminary studies [9,21], the standard linear model (1.2) was the basis for the extensions of the infinitesimal biphasic model incorporating intrinsic viscoelasticity. Consequently, any non-linear laws we develop should be reducible to (1.2) under the assumptions of infinitesimal strain and uni-axial motion.

**3.2.5. Symmetry and isotropy.** For most soft tissues, there is no moment-of-momentum supply and the balance of angular momentum implies symmetry of the Cauchy stress tensor (and hence the material stress tensor). Although soft tissues can exhibit a variety of material anisotropy, the characterization of both viscoelasticity and anisotropy in relation to other physical mechanisms in soft tissues is still in its infancy. For this initial formulation, we then take the liberty of assuming that the material can be taken to be isotropic. The constitutive laws for the second Piola-Kirchhoff stress and free energy must then be invariant under time-independent rigid rotations of the material frame given by $\mathbf{X}' = \mathbf{Q}\mathbf{X}$, where $\mathbf{Q}$ is a fixed proper rigid rotation.

**4. Formulation.** We are now in a position to formulate a set of constitutive laws which are of the form (3.2) and satisfy the requirements stated above.

**4.1. Derivation of relations satisfying dissipation.** Substitution of the constitutive functions (3.2) into the inequality (3.3) gives:

$$tr\left\{\left(-\rho_0\frac{\partial\bar{\Psi}}{\partial\mathbf{C}}+\frac{1}{2}\bar{\mathbf{S}}\right)\dot{\mathbf{C}}-\left(\rho_0\frac{\partial\bar{\Psi}}{\partial\mathbf{A}}\right)\bar{\mathbf{K}}\right\}\geq 0 \tag{4.1}$$

Now (4.1) must certainly hold in this case where $\gamma=\gamma^0$, $\alpha=\alpha^0$ so that:

$$tr\left\{\left(-\rho_0\frac{\partial\bar{\Psi}^s}{\partial\mathbf{C}}+\frac{1}{2}\bar{\mathbf{S}}^s\right)\dot{\mathbf{C}}\right\}\geq 0 \tag{4.2}$$

Since $\bar{\mathbf{S}}^s$, $\bar{\Psi}^s$ are independent of $\dot{\mathbf{C}}$, $\gamma$ and $\alpha$, (4.2) can only hold true in general if:

$$\bar{\mathbf{S}}^s(\mathbf{C})=2\rho_0\frac{\partial\bar{\Psi}^s}{\partial\mathbf{C}} \tag{4.3}$$

As a result, it is apparent that the stored stress is hyperelastic.

Substituting (4.3) into (3.2), and the resulting stress and energy back into (4.1) gives:

$$tr\left\{\left(\frac{1}{2}\bar{\mathbf{S}}^d\right)\dot{\mathbf{C}}-\left(\rho_0\frac{\partial\bar{\Psi}^d}{\partial\mathbf{A}}\right)\bar{\mathbf{K}}\right\}\geq 0 \tag{4.4}$$

We now wish to determine additional constitutive relations involving the dissipative stress function $\bar{\mathbf{S}}^d$, the dissipative energy function $\bar{\Psi}^d$ and the evolution function $\bar{\mathbf{K}}$ that satisfy (4.4). Consistency with the relation between the dissipative stress and energy for the standard linear model (1.2) (Figure 1) can be achieved via the choice:

$$\bar{\mathbf{S}}^d(\mathbf{A};\gamma)=2\rho_0\frac{\partial\bar{\Psi}^d}{\partial\mathbf{A}} \tag{4.5}$$

which reduces (4.4) to:

$$tr\left\{\rho_0\frac{\partial\bar{\Psi}^d}{\partial\mathbf{A}}\left(\dot{\mathbf{C}}-\bar{\mathbf{K}}(\dot{\mathbf{C}},\mathbf{A})\right)\right\}\geq 0 \tag{4.6}$$

We are now in a position to choose a relation between the evolution function $\bar{\mathbf{K}}$ and the dissipative energy function $\bar{\Psi}^d$ that satisfies (4.6). A natural choice is given by:

$$\dot{\mathbf{A}}=\bar{\mathbf{K}}(\dot{\mathbf{C}},\mathbf{A})=\dot{\mathbf{C}}-2\rho_0\mathcal{K}(\dot{\mathbf{C}},\mathbf{A})\frac{\partial\Psi^d}{\partial\mathbf{A}} \tag{4.7}$$

where $\mathcal{K}$ is a positive semi-definite fourth-rank tensor called the *dissipation tensor*.

It then follows that equations (3.2),(4.3), (4.5) and (4.7) form a set of constitutive relations for a viscoelastic solid in finite strain that satisfy the requirement of dissipation (§3.2.1). However, they are incomplete in that we still require expressions relating the free energies in (4.3) and (4.5) to the kinematics of motion. Furthermore, until such expressions are chosen, we cannot properly address the additional requirements of §3.2.2-3.2.5. Motivated by this, we now present a *finite linear model* as an illustration of a complete set of three dimensional viscoelastic constitutive laws that are valid in finite strain and satisfy the requirements of §3.2.1-3.2.5. In this case, we will see that the relations (3.2),(4.3), (4.5) and (4.7) can be combined to yield a single rate-type constitutive law.

**4.2. Formulation of a finite linear model.** To begin, we present a general form of the constitutive relation for the stored stress given by (4.3). Since the stored stress and energy are due to the hyperelastic behavior of an isotropic solid, the representation theorems [24] can be used to write $\bar{\Psi}^s(\mathbf{C}) = \tilde{\Psi}^s(I, II, III)$. In this case, the constitutive relation (4.3) gives:

$$(4.8)\quad \bar{\mathbf{S}}^s(\mathbf{C}) = 2\rho_0\left(a_0\mathbf{I} + a_1\mathbf{C} + a_2\mathbf{C}^2\right) \qquad \sigma^s = \frac{2\rho_0}{J}\left(a_0\mathbf{b} + a_1\mathbf{b}^2 + a_2\mathbf{b}^3\right)$$

where:

$$(4.9)\quad a_0 = \left(\frac{\partial}{\partial I} + I\frac{\partial}{\partial II} + II\frac{\partial}{\partial III}\right)\tilde{\Psi}^s \qquad a_1 = \left(-\frac{\partial}{\partial II} - I\frac{\partial}{\partial III}\right)\tilde{\Psi}^s \qquad a_2 = \frac{\partial\tilde{\Psi}^s}{\partial III}$$

In (4.8)-(4.9), $I, II, III$ are the three *principal invariants* of the material strain tensor $\mathbf{C}$. They are related to the *principal stretches* $\lambda_1, \lambda_2, \lambda_3$ by:

$$(4.10)\quad I = \lambda_1^2+\lambda_2^2+\lambda_3^2 \qquad II = \lambda_1^2\lambda_2^2+\lambda_1^2\lambda_3^2+\lambda_2^2\lambda_3^2 \qquad III = \lambda_1^2\lambda_2^2\lambda_3^2 = J^2$$

and the *principal stored stresses* $\sigma_1^s, \sigma_2^s, \sigma_3^s$ act along the eigendirections of the *left Cauchy-Green deformation tensor* $\mathbf{b}$. The specific form of $\tilde{\Psi}^s$ depends on the hyperelastic model chosen. As a result of our additive decomposition (3.2), we can leave the choice of the hyperelastic part of the free energy open. However, we will require that the stored energy function $\tilde{\Psi}^s$ be non-negative for all motions so that:

$$(4.11)\qquad \tilde{\Psi}^s(I, II, III) \geq 0 \qquad \text{for all admissible processes}$$

In addition, to ensure that the equilibrium stress does not exhibit behavior implausible for an elastic material, we will require that the principal stored stresses satisfy two fundamental inequalities. These are the *Baker-Ericksen (B-E) inequality* and the *tension-extension inequality* [24] which, using (4.8)- (4.9), can be written (respectively) as:

$$(4.12)\quad (i)\ \frac{\partial\tilde{\Psi}^s}{\partial I}+\lambda_i^2\frac{\partial\tilde{\Psi}^s}{\partial II}>0 \qquad (ii)\ \frac{\partial\sigma_i^s}{\partial\lambda_i}>0 \quad (\text{no sum on } i) \quad i=1,2,3$$

In the following analysis, we may then assume that the stored stress is of the general form (4.8) and satisfies the inequalities (4.11)-(4.12).

We now consider the constitutive relation (4.5) for the dissipative stress $\bar{\mathbf{S}}^d$. Inherent to the additive decomposition (3.2) is an assumption that the tensor $\mathbf{A}$ quantifies an internal mechanism that contributes to the total stress during only the transient response to any prescribed loading or deformation. As a result, we have stated the requirement (3.4) in §3.2.2 that the internal material tensor assume the constant value $\mathbf{A} = \mathbf{A}^{eq}$ in any equilibrium state. With regards to characterizing the viscoelasticity of soft tissues, one of the primary goals is to establish the relative contribution of viscoelasticity in relation to other physical mechanisms. More precisely, there is a need to quantify the viscoelastic effect as a perturbation of the hyperelastic model. Motivated by this, we can expand the dissipative stress function in powers of the components of $\mathbf{A} - \mathbf{A}^{eq}$ and take the first term (linear) approximation. If the additional requirements of symmetry and isotropy (§3.2.5) are introduced, then the representation theorems in [24] can be used to write:

$$(4.13)\qquad \bar{\mathbf{S}}^d(\mathbf{A};\gamma) = 2\rho_0\Big(\beta_1 tr(\mathbf{A}-\mathbf{A}^{eq})\mathbf{I} + \beta_2(\mathbf{A}-\mathbf{A}^{eq})\Big)$$

Due to the linear nature of (4.13), we will call the set of constitutive laws that result a *finite linear model.* Using the constitutive relation (4.5), it is straightforward to show that (4.13) is derivable from the dissipative free energy function:

$$(4.14)\qquad \bar{\Psi}^d(\mathbf{A};\alpha) = \frac{\beta_1}{2}\Big(tr(\mathbf{A}-\mathbf{A}^{eq})\Big)^2 + \frac{\beta_2}{2}\, tr\Big((\mathbf{A}-\mathbf{A}^{eq})^2\Big)$$

The sets of viscoelastic material constants described in §3.2.3 are then given by $\gamma = \alpha = (\beta_1, \beta_2)$.

To complete our set of viscoelastic constitutive laws, we must choose the dissipation tensor $\mathcal{K}(\dot{\mathbf{C}}, \mathbf{A})$ in the evolution function (4.7). In the general case, the dissipation tensor consists of 81 functions of the components of $\dot{\mathbf{C}}$ and $\mathbf{A}$ and must satisfy a condition of positive semi-definiteness. The requirement of symmetry (§3.2.5) reduces the number of functions to 27, while the requirement of isotropy (§3.2.5) implies that the components are functions of only the three principal invariants of the tensors $\dot{\mathbf{C}}$ and $\mathbf{A}$. However, for the purposes of our finite linear model, we wish to keep the number of viscoelastic material constants to a minimum. Consequently, we will make a major simplification by assuming that the dissipation tensor is given by a single dissipation constant $\kappa$ so that, in component form:

$$(4.15)\ \mathcal{K}_{IJKL} = \frac{1}{\kappa}\delta_{IK}\delta_{JL} \qquad \text{where:} \quad \delta_{MN} = \begin{cases} 1 & \text{if } M = N \\ 0 & \text{otherwise} \end{cases} \qquad \kappa \geq 0$$

Substitution of (4.15) and (4.13) into (4.7) reduces our evolution equation for $\mathbf{A}$ to:

$$\dot{\mathbf{A}} = \dot{\mathbf{C}} - \frac{2\rho_0}{\kappa}\Big(\beta_1 tr(\mathbf{A}-\mathbf{A}^{eq})\mathbf{I} + \beta_2(\mathbf{A}-\mathbf{A}^{eq})\Big) \tag{4.16}$$

It is then apparent that equations (3.2), (4.8), (4.13) and (4.16) form a complete set of constitutive laws to be called a finite linear model. We can eliminate the internal material tensor $\mathbf{A}$ and combine all four equations to obtain a single rate-type constitutive law. If we first solve (4.13) for $\mathbf{A}$, substitute the result into the evolution function (4.16), and use the additive stress decomposition in (3.2), then we obtain the following rate-type law:

$$\begin{aligned}\mathbf{S} - \chi_1 tr(\mathbf{S})\mathbf{I} + \tau_2^S\dot{\mathbf{S}} &= \bar{\mathbf{S}}^s(\mathbf{C}) - \chi_1 tr(\bar{\mathbf{S}}^s(\mathbf{C}))\mathbf{I}\\ &\quad + H_A\tau_1^C\left(\left\{1 + \frac{\chi_2}{H_A}\frac{\partial\bar{\mathbf{S}}^s(\mathbf{C})}{\partial\mathbf{C}}\right\}\dot{\mathbf{C}} - \chi_1 tr(\dot{\mathbf{C}})\mathbf{I}\right)\end{aligned} \tag{4.17}$$

where $\chi_1 = \tau_1^S/(3\tau_1^S + \tau_2^S)$, $\chi_2 = \tau_2^S/\tau_1^C$ and the relaxation time constants $\tau_1^S, \tau_2^S, \tau_1^C$ are related to the viscoelastic material constants $\beta_1, \beta_2$ and $\kappa$ via:

$$\tau_1^S = -\frac{\beta_1\kappa}{2\rho_0\beta_2(3\beta_1+\beta_2)} \qquad \tau_2^S = \frac{\kappa}{2\rho_0\beta_2} \qquad \tau_1^C = \frac{\kappa}{H_A} \tag{4.18}$$

In particular, we see that (4.17) is in the class of rate-type laws (1.3) and allows for a retention of the dependence of the coefficient of the strain rate $\dot{\mathbf{E}}$ on the strain $\mathbf{E}$. This is in contrast to the analysis of [19], in which this coefficient was taken to be constant.

If we assume that the material is unstrained at $t = -\infty$, then the law (4.17) can also be written in integral form as:

$$\begin{aligned}\mathbf{S} &= \mathbf{S}^s(\mathbf{C}) - 2\rho_0\beta_2\int_0^\infty e^{-s/\tau_2^S}\frac{\partial}{\partial s}\tilde{\mathbf{C}}(\mathbf{X}, t-s)ds\\ &\quad + 2\rho_0\frac{3\beta_1+\beta_2}{3}\int_0^\infty e^{-s/(3\tau_1^S+\tau_2^S)}\frac{\partial}{\partial s}tr(\mathbf{C}(\mathbf{X}, t-s))\mathbf{I}ds\end{aligned} \tag{4.19}$$

provided the following conditions, necessary for convergence of the integrals, hold:

$$3\tau_1^S + \tau_2^S > 0 \quad \text{and} \quad \tau_2^S > 0 \tag{4.20}$$

It may be observed that the two relaxation times $\tau_1^S, \tau_2^S$ are independent in the sense that $\tau_2^S$ is associated with the relaxation of the *deviatoric strain* tensor $\tilde{\mathbf{C}} \equiv \mathbf{C} - 1/3tr(\mathbf{C})\mathbf{I}$, while the sum $3\tau_1^S + \tau_2^S$ is associated with the relaxation of the *spherical strain* tensor $tr(\mathbf{C})\mathbf{I}$.

**4.3. Satisfaction of the constitutive requirements.** We can now demonstrate that the finite linear model given by (4.17) satisfies the requirements of §3.2.1-3.2.5. By the derivation in §4.1, the requirement of dissipation (§3.2.1) has already been satisfied. The choices (4.8), (4.13) and (4.16) also ensure that the requirements of symmetry and isotropy (§3.2.5) are satisfied. By (4.13), the requirement (3.7) of §3.2.3 is also satisfied as the choice $(\beta_1, \beta_2) = (0, 0)$ reduces the stress and energy in (3.2) to those of hyperelasticity. It then remains to examine the considerations of energy in §3.2.2 and the requirement of consistency with the standard linear model under infinitesimal strain (§3.2.4).

We first examine the energy considerations given by conditions (3.4)-(3.6). Condition (3.5) states that the total free energy should remain non-negative. Since, by (4.11), we assumed that the stored energy function remains non-negative, it suffices to show the the dissipative energy function in (4.14) does the same. If we expand the dissipative energy in the components of the symmetric tensor $\mathbf{B} \equiv \mathbf{A} - \mathbf{A}^{eq}$, then the requirement of non-negative energy becomes:

$$\frac{\beta_1}{2}(tr(\mathbf{B}))^2 + \frac{\beta_2}{2}tr(\mathbf{B}^2) = \frac{\beta_2}{2}tr(\tilde{\mathbf{B}}^2) + \frac{3\beta_1 + \beta_2}{6}(tr(\mathbf{B}))^2 \geq 0 \tag{4.21}$$

Integration of the evolution function (4.16) yields the relation:

$$\begin{aligned} \mathbf{B} = \mathbf{C} - \frac{1}{3(3\tau_1^S + \tau_2^S)} \int_0^\infty & e^{-s/(3\tau_1^S+\tau_2^S)} tr(\mathbf{C}(\mathbf{X}, t-s))\mathbf{I} ds \\ & - \frac{1}{\tau_2^S} \int_0^\infty e^{s/\tau_2^S} \tilde{\mathbf{C}}(\mathbf{X}, t-s) ds \end{aligned} \tag{4.22}$$

From (4.22), it is apparent that the two terms in (4.21) can be varied independently. As a result, satisfaction of the inequality in (4.21) can then be guaranteed if the viscoelastic material constants satisfy:

$$\text{Either:} \quad 3\beta_1 + \beta_2 > 0 \quad \text{and} \quad \beta_2 > 0 \quad \text{or:} \quad \beta_1 = \beta_2 = 0 \tag{4.23}$$

These inequalities can then be used in combination with the constraint on the dissipation constant in (4.15) to show that relaxation time constants given by (4.18) must satisfy the inequalities (4.20). It is noted that the conditions (4.20) are not required in the elastic case as the coefficients $\beta_1$ and $\beta_2$ in (4.19) vanish identically. As a result, it is evident that the requirement of positive energy guarantees convergence of the integrals in (4.19).

Condition (3.4) states that the internal material tensor $\mathbf{A}$ must tend to the constant state $\mathbf{A}^{eq}$ whenever equilibrium ($\mathbf{C} \to \mathbf{C}_\infty, \dot{\mathbf{C}} \to \mathbf{0}$ as $t \to \infty$) is achieved. The representation (4.22) for the internal material tensor can be re-written as:

$$
\begin{aligned}
\mathbf{A}-\mathbf{A}^{eq} &= \frac{1}{3}e^{-t/(3\tau_1^S+\tau_2^S)}\int_{-\infty}^{t} e^{s/(3\tau_1^S+\tau_2^S)}\frac{\partial}{\partial s}tr(\mathbf{C}(\mathbf{X},s))\mathbf{I}ds \\
&\quad + e^{-t/\tau_2^S}\int_{-\infty}^{t} e^{s/\tau_2^S}\frac{\partial}{\partial s}\tilde{\mathbf{C}}(\mathbf{X},s)ds
\end{aligned}
\tag{4.24}
$$

where it has been assumed that, at $t=-\infty$, the material is unstrained and $\mathbf{A}=\mathbf{A}^{eq}$ . Assuming a steady state is achieved as $t\rightarrow\infty$, then $\dot{\mathbf{C}}\rightarrow\mathbf{0}$ and L'Hopital's rule may applied to (4.24) to show that the right hand side vanishes and hence $\mathbf{A}\rightarrow\mathbf{A}^{eq}$, so that condition (3.4) is satisfied. Satisfaction of condition (3.6), that the equilibrium state is hyperelastic, then follows from the additive decomposition (3.2) and (4.8) since, by (4.13), $\bar{\mathbf{S}}^d(\mathbf{A})\rightarrow\mathbf{0}$ as $\mathbf{A}\rightarrow\mathbf{A}^{eq}$.

It is then apparent that the considerations of energy given by conditions (3.4),(3.5) and (3.6) are all satisfied by our finite linear model. It should also be noted that these requirements are satisfied without having to specify the value of the equilibrium state $\mathbf{A}^{eq}$ for the internal material tensor. This allows for some flexibility in tailoring the definition of $\mathbf{A}$ to the internal dissipation mechanism appropriate for the specific material of interest.

**4.4. Three dimensional infinitesimal model.** It remains to show that the finite linear model satisfies the requirement (§3.2.4) of consistency with the one dimensional standard linear model (1.2). We can demonstrate this through a simplification of the finite linear model under the assumption of infinitesimal strains, in which case $\mathbf{C}\sim\epsilon$, $\mathbf{S}\sim\sigma$. Here, $\epsilon$ is the *infinitesimal strain tensor* defined by $\epsilon=1/2(\nabla\mathbf{u}+\nabla\mathbf{u}^t)$, where the *relative displacement* is $\mathbf{u}=\mathbf{x}-\mathbf{X}$. In this case, to leading order, the constitutive laws (4.8) and (4.17) reduce to the single rate-type law:

$$
\sigma+\tau_1^S tr(\dot{\sigma})\mathbf{i}+\tau_2^S\dot{\sigma}=H_A(\bar{\lambda}^s tr(\epsilon)\mathbf{i}+2\bar{\mu}^s\epsilon+\tau_1^\epsilon tr(\dot{\epsilon})\mathbf{i}+\tau_2^\epsilon\dot{\epsilon})
\tag{4.25}
$$

where $H_A=\lambda^s+2\mu^s$, $\bar{\lambda}^s=\lambda^s/H_A$, $\bar{\mu}^s=\mu^s/H_A$ and:

$$
\tau_1^\epsilon=(3\bar{\lambda}^s+2\bar{\mu}^s)\tau_1^S+\bar{\lambda}^s\tau_2^S \qquad\qquad \tau_2^\epsilon=2\bar{\mu}^s\tau_2^S+\tau_1^C
\tag{4.26}
$$

It is apparent that (4.25) can be postulated *a priori* as the simplest linear isotropic relationship between $\sigma,\dot{\sigma},\epsilon,\dot{\epsilon}$ resulting in the four viscoelastic material constants $\tau_1^S,\tau_2^S,\tau_1^\epsilon,\tau_2^\epsilon$. However, closer examination of (4.26) indicates that there are really only three independent viscoelastic constants. As a result of our thermodynamic considerations (4.23) (and hence (4.20)) and the elastic *C-inequalities* [24], the four time constants in (4.25) are constrained by:

$$
\tau_2^S>0 \quad 3\tau_1^S+\tau_2^S>0 \quad \tau_2^\epsilon>2\bar{\mu}^s\tau_2^S \quad \tau_1^\epsilon=(3\bar{\lambda}^s+2\bar{\mu}^s)\tau_1^S+\bar{\lambda}^s\tau_2^S
\tag{4.27}
$$

with the additional (elastic) state given by $\tau_1^S=\tau_2^S=0=\tau_1^\epsilon=\tau_2^\epsilon$.

Under the assumption of a uni-axial relative displacement $u(z,t)$ along a single principal axis with coordinate $z$, the relevant components of the infinitesimal strain and stress tensors are $\frac{\partial u}{\partial z}$ and $\sigma_{zz}$ (respectively) which satisfy:

$$\sigma_{zz} + \tau_1^S tr(\dot{\sigma}) + \tau_2^S \dot{\sigma}_{zz} = H_A \left( \frac{\partial u}{\partial z} + (\tau_1^\epsilon + \tau_2^\epsilon) \frac{\partial^2 u}{\partial z \partial t} \right) \tag{4.28}$$

To obtain the law (1.2), two possibilities are to take $\tau_1^S = 0$, or to assume that the principal stress rates along the other two axes are negligible. The first choice is consistent with an assumption that the relaxation functions associated with spherical and deviatoric strain in (4.19) are the same. For studies of articular cartilage in infinitesimal strain, such an assumption has been employed in [20,21]. The second choice is appropriate for studies of soft tissues in a configuration of uni-axial confined compression. With these assumptions, (4.28) reduces to the form (1.2) with either $\tau_0^\sigma = \tau_2^S$ or $\tau_0^\sigma = \tau_1^S + \tau_2^S$, and $\tau_0^u = \tau_2^\epsilon + \tau_1^\epsilon$. It is then apparent that the last of our constitutive requirements (§3.2.4) is satisfied by the finite linear model.

**5. Stress response to simple deformations.** We now examine the transient stress response of the finite linear model in configurations involving simple monotonic deformations. In these deformations, the material is unstrained for $t < 0$, equilibrium is achieved as $t \to \infty$ and it is reasonable to expect that, for all $t > 0$, the relevant component of Cauchy stress in the direction of the motion be directed in the same sense as the motion. As will be shown below, the finite linear model does satisfy this interpretation.

**5.1. Simple monotonic compression/extension.** First, we define a *simple monotonic compression* (*extension*) as a time dependent analogue of a simple compression (extension) for an elastic material. Let the function $f(t) > 0$ have the following properties:

$$(i)\ f(t) = 1 \ \text{ for } t \le 0 \qquad (ii)\ \lim_{t\to\infty} f(t) = f_\infty \qquad (iii)\ sgn(f'(t)) = n \tag{5.1}$$

where either $n = -1$ or $n = 1$ (exclusively) and $sgn(x) \equiv x/|x|$. In Cartesian coordinates, the following class of deformations may then be defined:

$$x^1 = f(t)X^1 \qquad x^2 = X^2 \qquad x^3 = X^3 \tag{5.2}$$

If $n = -1$ and $0 < f_\infty < 1$, (5.2) is called a *simple monotonic compression* while if $n = 1$ and $f_\infty > 1$ it will be called a *simple monotonic extension*. The deformation (5.2) may be interpreted as the time-dependent compression or extension of a unit cube along one principal axis (Figure 2(a)). Since the compression (extension) is monotone, it is reasonable to require that the component of Cauchy stress normal to the vertical face in motion be directed in the same sense as the motion so that:

$$sgn(\sigma_{11}) = n \tag{5.3}$$

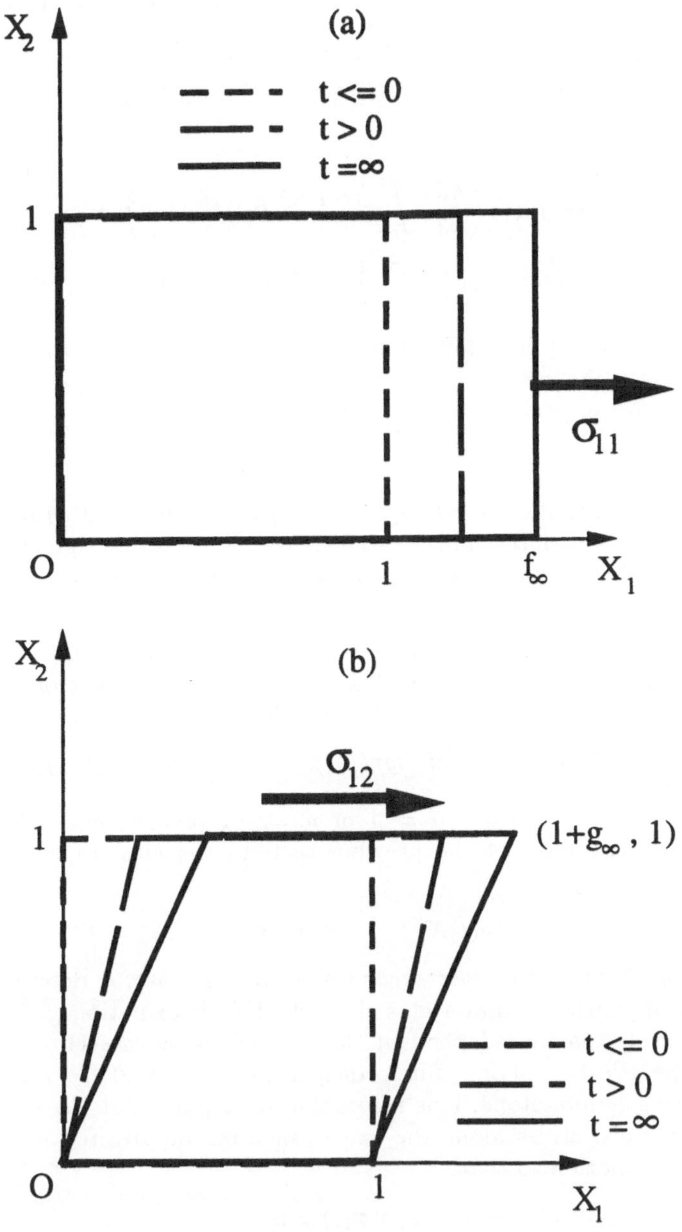


FIG. 2. *Deformation of a unit cross-section under simple motions: (a) simple monotonic extension ($n = 1$, $f_\infty > 1$) (b) simple monotonic shear (the case $n = 1$ is shown).*

Under the simple monotonic compression/extension given by (5.2), it can be shown, using (4.8),(4.19) that $\sigma_{11} = \sigma_{11}^s + \sigma_{11}^d$ where:

$$\sigma_{11}^s = \frac{2\rho_0}{J}(a_0 f^2(t) + a_1 f^4(t) + a_2 f^6(t)) \tag{5.4}$$

and:

$$\begin{aligned}\sigma_{11}^d = &\frac{2\rho_0}{J} f^2(t)\left(\frac{4\beta_2}{3}\int_0^t e^{(s-t)/\tau_2^S} f(s)f'(s)ds\right)\\ &+ \frac{2\rho_0}{J} f^2(t)\left(2\frac{3\beta_1+\beta_2}{3}\int_0^t e^{(s-t)/(3\tau_1^S+\tau_2^S)} f(s)f'(s)ds\right)\end{aligned} \tag{5.5}$$

Since $\mathbf{C}$ is diagonal, it follows that $\sigma_{11}^s = \sigma_1^s$, $\lambda_1 = f(t)$ and the inequality (4.12)(ii) then reduces to:

$$\frac{\partial \sigma_{11}^s}{\partial \lambda_1} = \frac{\partial \sigma_{11}^s}{\partial t}\left(\frac{\partial \lambda_1}{\partial t}\right)^{-1} = \frac{\partial \sigma_{11}^s}{\partial t}(f'(t))^{-1} > 0 \tag{5.6}$$

If the solid is initially stress free, then (5.1)(iii) implies that $sgn(\sigma_{11}^s) = n$. Now by (5.1) both integrands in (5.5) maintain the same sign so that the inequalities (4.23) guarantee that $sgn(\sigma_{11}^d) = n$ and hence that $sgn(\sigma_{11}) = n$.

**5.2. Simple monotonic shear.** Similarly, define a *simple monotonic shear* as the time-dependent analogue of a simple shearing of an elastic material. Let the function $g(t)$ have the following properties:

$$(i)\ g(t) = 0 \text{ for } t \le 0 \quad (ii)\ sgn(g(t), g'(t)) = n \quad (iii)\ \lim_{t\to\infty} g(t) = ng_\infty \tag{5.7}$$

where $g_\infty > 0$ and either $n = 1$ or $n = -1$ (exclusively). Using the Cartesian coordinates of the previous section, the class of deformations defined by:

$$x^1 = X^1 + g(t)X^2 \qquad x^2 = X^2 \qquad x^3 = X^3 \tag{5.8}$$

is called a *simple monotonic shear* where, in this case, $n$ determines the direction in which the material is sheared. The deformation (5.8) may be interpreted as the time-dependent shearing of one cross-section of a unit cube in the direction of one of its principal axes (Figure 2(b)). Again, since the shearing is monotone, it is reasonable to require that the component of Cauchy shear stress along the axis of shearing be directed in the same sense as the motion so that:

$$sgn(\sigma_{12}) = n \tag{5.9}$$

Using (5.8) and (4.10) it can be shown that the invariants and principal stretches are then given by:

$$\begin{aligned} &I = 3 + g^2(t) \qquad II = 3 + g^2(t) \qquad III = 1\\ &\lambda_3 = 1 \qquad \lambda_1, \lambda_2 \text{ roots of: } \ z^2 - (2 + g^2(t))z + 1 = 0\end{aligned} \tag{5.10}$$

and using (5.10) and (4.8) in (4.19) gives $\sigma_{12} = \sigma_{12}^s + \sigma_{12}^d$ where:

$$\sigma_{12}^s = \frac{2\rho_0}{J} g(t) \left( \frac{\partial \tilde{\Psi}^s}{\partial I} + \frac{\partial \tilde{\Psi}^s}{\partial II} \right) \tag{5.11}$$

and:

$$\begin{aligned}\sigma_{12}^d = {} & \frac{2\rho_0}{J} \left( \beta_2 \int_0^t e^{(s-t)/\tau_2^S} \left(1 + \frac{4}{3} g(t) g(s)\right) g'(s) ds \right) \\ & + \frac{2\rho_0}{J} \left( 2 \frac{3\beta_1 + \beta_2}{3} g(t) \int_0^t e^{(s-t)/(3\tau_1^S + \tau_2^S)} g(s) g'(s) ds \right)\end{aligned} \tag{5.12}$$

By substituting $\lambda_3 = 1$ (from (5.10)) into the B-E inequality (4.12)(i), it then follows, from (5.7)(ii), that $sgn(\sigma_{12}^s) = sgn(g(t)) = n$. Again, by (5.7), the integrands in (5.12) maintain the same sign and the inequalities (4.23) guarantee that $sgn(\sigma_{12}^d) = sgn(g(t)) = n$ and hence that $sgn(\sigma_{12}) = n$.

**5.3. Simple monotonic torsional shear.** An analogy to the simple shearing of the previous section can also be examined in torsion. We use the cylindrical polar coordinates $(X^1, X^2, X^3) \equiv (R, \Phi, Z)$ and $(x^1, x^2, x^3) \equiv (r, \phi, z)$. The class of deformations defined by:

$$x^1 = X^1 \qquad x^2 = X^2 + g(t) X^3 \qquad x^3 = X^3 \tag{5.13}$$

is called a *simple monotonic torsional shear*. Again, it is reasonable to expect that, on axial cross sections ($X^3 = Z = const.$), the tangential component of Cauchy shear stress be directed in the same sense as the motion so that:

$$sgn(\sigma^{23}) = n \tag{5.14}$$

The invariants and principal stretches are given by:

$$\begin{aligned} & I = 3 + (X^1)^2 g^2(t) \qquad II = 3 + (X^1)^2 g^2(t) \qquad III = 1 \\ & \lambda_3 = 1 \qquad \lambda_1, \lambda_2 \text{ roots of: } \; z^2 - \left(2 + (X^1)^2 g^2(t)\right) z + 1 = 0 \end{aligned} \tag{5.15}$$

and using (5.15) and (4.8) in (4.19) gives $\sigma^{23} = (\sigma^s)^{23} + (\sigma^d)^{23}$ where:

$$(\sigma^s)^{23} = \frac{2\rho_0}{J} g(t) \left( \frac{\partial \tilde{\Psi}^s}{\partial I} + \frac{\partial \tilde{\Psi}^s}{\partial II} \right) \tag{5.16}$$

and:

$$\begin{aligned}(\sigma^d)^{23} = {} & \frac{2\rho_0}{J} \left( \beta_2 \int_0^t e^{(s-t)/\tau_2^S} \left(1 + \frac{4}{3} (X^1)^2 g(t) g(s)\right) g'(s) ds \right) \\ & + \frac{2\rho_0}{J} \left( 2 \frac{3\beta_1 + \beta_2}{3} (X^1)^2 g(t) \int_0^t e^{(s-t)/(3\tau_1^S + \tau_2^S)} g(s) g'(s) ds \right)\end{aligned} \tag{5.17}$$

Using the argument of the previous section, substitution of $\lambda_3 = 1$ (from (5.15)) into the B-E inequality (4.12)(i) gives, by (5.7)(ii), $sgn((\sigma^s)^{23}) = sgn(g(t)) = n$. The integrands in (5.17) again maintain the same sign so that the inequalities (4.23) guarantee that $sgn((\sigma^d)^{23}) = sgn(g(t)) = n$ and hence that $sgn(\sigma^{23}) = n$.

**6. Discussion.** The rate-type finite linear law (4.17) provides a model for finite strain viscoelasticity that satisfies objectivity, dissipation and some additional requirements that are appropriate for many biological soft tissues. This formulation has the advantage of introducing only three viscoelastic material constants and allowing for an arbitrary hyperelastic equilibrium response. Consequently, the law (4.17) can be used to assess the relative contribution of the intrinsic viscoelastic effect as a perturbation of existing non-linear biphasic models [8]. The material (Lagrangian) description that we have employed also has many advantages in developing weak formulations of the biphasic-viscoelastic model for use in finite element simulations of complex three dimensional joint mechanics.

It is important to note that the finite linear model is a first approximation in which the internal material tensor can be eliminated from the formulation. Whether such a model is sufficiently non-linear for modeling the intrinsic viscoelasticity of soft tissues remains an open question. Higher order approximations, in which it may not be possible to eliminate the internal material tensor, will require models for the kinematics of the internal dissipation mechanism. Through a combination of theoretical and experimental studies of both the macroscopic and microscopic response of soft tissues to transient loading and deformation, it is our hope to answer these questions and, where appropriate, develop extended models.

## REFERENCES


[1] BIOT, M.A. (1972), *Theory of finite deformations of porous solids*, *J. Applied Phys.* **33**, 1482–1498.

[2] BOWEN, R.M. (1980), *Incompressible porous media models by use of the theory of mixtures*, *Int. J. Engng Sci.* **18**, 1129–1148.

[3] COLEMAN, B.D. AND GURTIN, M.E. (1967), *Thermodynamics with internal state variables*, *J. Chem. Phys.* **47**, 597–613.

[4] COLEMAN, B.D. AND NOLL, W. (1967), *Foundations of linear viscoelasticity*, *Rev. Modern Phys.* **33**, 239–249.

[5] FABRIZIO, M. AND MORRO, A. (1992), *Mathematical Problems in Linear Viscoelasticity*, SIAM, Philadelphia.

[6] FUNG, Y.C. (1993), *Biomechanics: Mechanical Properties of Living Tissues*, 2nd ed., Springer-Verlag, New York.

[7] HAYES, W.C. AND BODINE, A.J. (1978), *Flow-independent viscoelastic properties of articular cartilage matrix*, *J. Biomech.* **11**, 407–420.

[8] HOLMES, M.H. (1986), *Finite deformation of soft tissue: Analysis of a mixture model in uni-axial compression*, *J. Biomech. Engng* **108**, 372–381.

[9] HOLMES, M.H. AND HAIDER, M.A. (1994), *The role of matrix viscoelasticity and the fluid flow in the compressive behavior of cartilage*, *Proc. Second World Congree of Biomechanics* v.II, 29, Stichting, Nijmegen.

[10] HOLMES, M.H. AND MOW, V.C. (1990), *The nonlinear characteristics of soft gels and hydrated connective tissues in ultrafiltration*, *J. Biomech.* **23**, 1145–1156.

[11] LAI, W.M, MOW, V.C. AND ROTH, V. (1981), *Effects of of a nonlinear strain-dependent permeability and rate of compression on the stress behavior of articular cartilage*, *J. Biomech. Engng* **103**, 61–66.

[12] LETALLEC, P. AND RAHIER, C. (1994), *Numerical models of steady rolling for nonlinear viscoelastic structures in finite deformations*, *IJNME* **37**, 1159–1186.

[13] LETALLEC, P., RAHIER, C. AND KAISS, A. (1993), *Three dimensional incompressible viscoelasticity in large strains: Formulation and numerical approximation*, *Comput. Methods Appl. Mech. Eng.* **109**, 233–258.

[14] LUBLINER, J. (1985), *A model of rubber viscoelasticity*, *Mech. Res. Comm.* **12**, 93–99.

[15] MALVERN, L.E. (1969), *Introduction to the Mechanics of a Continuous Medium*, Prentice-Hall, Englewood Cliffs, New Jersey.

[16] MARSDEN, J.E. AND HUGHES, T.J.R. (1994), *Mathematical Foundations of Elasticity*, Dover, New York.

[17] MOW, V.C., HOLMES, M.H. AND LAI, W.M. (1984), *Fluid transport and mechanical properties of articular cartilage: a review*, *J. Biomech.* **17**, 377–394.

[18] PASSMAN, S.L., NUNZIATO, J.W. AND WALSH, E.K. (1984), *A theory of multiphase mixtures*, in *Raional Thermodynamics*, Springer-Verlag, New York.

[19] PODIO-GUIDUGLI, P. AND SULICIU, I. (1984), *On rate-type viscoelasticity and the second law of thermodynamics*, *Int. J. Non-Linear Mechanics* **19**, 545–564.

[20] SETTON, L.A., GU, W.Y., MULLER, F.J., PITA, J.C. AND MOW, V.C. (1992), *Changes in the intrinsic shear behavior of articular cartilage with joint disuse*, *Trans. Orthop. Res. Soc.* **17**, 209.

[21] SETTON, L.A., ZHU, W. AND MOW, V.C. (1993), *The biphasic poroviscoelastic behavior of articular cartilage: Role of the surface zone in governing the compressive behavior*, *J. Biomech.* **26**, 581–592.

[22] SIMO, J.C. (1987), *On a fully three-dimensional finite-strain viscoelastic damage model: Formulation and computational aspects*, *Comput. Methods Appl. Mech. Eng.* **60**, 153–173.

[23] TERZAGHI, K. (1943), *Theoretical Soil Mechanics*, John Wiley and Sons, New York.

[24] TRUESDELL, C. AND NOLL, W. (1992), *The Non-Linear Field Theories of Mechanics*, Springer-Verlag, New York.

[25] TRUESDELL, C. AND TOUPIN, R. (1960), *The classical field theories*, in *Handbuch der Physik* (S. Flügge, Ed.), Springer-Verlag, Berlin, 226–793.

# DRAG IN A POROUS MEDIUM: AN EXAMPLE OF THE USE OF ENSEMBLE AVERAGED HYDRODYNAMIC POTENTIALS

MARK F. HURWITZ*

**Abstract.** The drag in a porous medium, composed of randomly positioned identical rigid spheres, is calculated as an example of a general framework for the analysis of multiphase flows. The field variables are treated as sums of known fluid fields, independent of the presence of the solid phase, and disturbance fields which account for the effect of the solid phase on the fluid phase. The problem of calculating the ensemble average interaction force is transformed to one related to singular solutions of Stokes equations by representing the disturbance fields in terms of hydrodynamic potentials. Approximate solutions of the problem of a point force outside of two spheres are used to directly calculate the flow resistance in a porous medium. The result is very accurate for solid volume fractions less than 0.5. Because of the inaccuracy of the approximation of the far field interactions among the rigid spheres, the solution is less accurate for more densely packed porous media.

**Key words.** porous flow hydrodynamic potential ensemble average.

**1. Introduction.** We present a calculation of the drag in a porous medium composed of randomly positioned identical rigid spheres. A more detailed exposition of this work is presented in our doctoral dissertation [1]. Following Drew [2], ensemble averaged fields are defined and the equations that they satisfy are determined. As in any averaging scheme, the equations contain open terms that are not represented as combinations of the averaged quantities and their derivatives. For example, the ensemble averaged momentum balance for the fluid phase is

$$\partial_t(\rho_f \dot{\mathbf{x}}_f) + \nabla \cdot (\rho_f \dot{\mathbf{x}}_f \dot{\mathbf{x}}_f - \mathbf{T}_f) = \rho_f \mathbf{B} + \mathbf{M}_f \tag{1.1}$$

in which $\rho_f$ is the average fluid density, $\dot{\mathbf{x}}_f$ is the average velocity, and $\mathbf{B}$ is any externally applied acceleration, such as gravity, that may be present. The open terms in this equation are the average fluid stress, $\mathbf{T}_f$ and the interaction force density $\mathbf{M}_f$.

To determine the open terms, we decompose the flow field into a background part and a disturbance part. The background is the flow which would exist if the rigid spheres were not present. That is, the solution of the Navier-Stokes equations which satisfy the same boundary conditions as the mixture. The disturbance flow is the difference between the background and the actual flow field. It is then easily shown that the average fluid stress, $\mathbf{T}_f$, is

$$\begin{aligned} \mathbf{T}_f = (1-\phi)\Big(2\mu \,\mathrm{SYM}\, \nabla \dot{\mathbf{x}}_f - \mathbf{1}\Big(P_\infty + \frac{1}{1-\phi}\langle H_f p\rangle\Big)\Big) + \boldsymbol{\tau}_f \\ + 2\mu\,\mathrm{SYM}\Big((\dot{\mathbf{x}}_f - \dot{\mathbf{x}}_\infty)\nabla(1-\phi) - \langle \mathbf{u}\nabla H_f\rangle\Big) \end{aligned} \tag{1.2}$$

* Pall Corporation, 3669 State Route 281, Cortland, New York 13045.

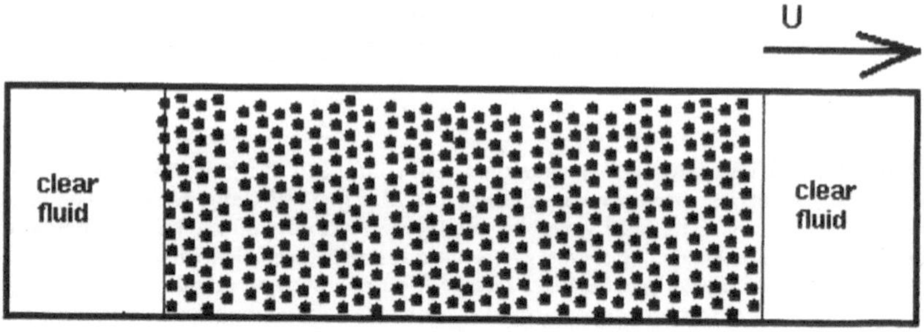


FIG. 1. *Model System for Drag Calculation.*

where $\dot{\mathbf{x}}_\infty$ and $P_\infty$ are the background velocity and pressure. The disturbance velocity and pressure are $\mathbf{u}$ and $p$, $\phi$ is the solid volume fraction, and $H_f$ is the fluid characteristic function which is zero everywhere except in the fluid, where it is one. The bracket pair $\langle\rangle$ is the ensemble average operator and $\boldsymbol{\tau}_f$ is the fluctuation stress.

The interaction force density $\mathbf{M}_f$ is given by

$$\mathbf{M}_f = \nabla\phi(\mathbf{x},t) \cdot^{\mathsf{T}} \mathbf{T}_\infty(\mathbf{x},t) + \int_{\mathbf{y}\in\partial\Omega} \delta(\mathbf{x}-\mathbf{y})\langle\mathbf{T}_d\rangle(\mathbf{y},t)\cdot \mathbf{da} - \tag{1.3}$$
$$-N\int_{|\mathbf{r}|=\sigma^+}\int_{\mathbf{w}_1}\int_{\mathbf{z}_1} q_1(\mathbf{x}-\mathbf{r},\mathbf{w}_1,\mathbf{z}_1,t)\langle\mathbf{T}_d\rangle_1(\mathbf{x},t\,|\mathbf{x}-\mathbf{r},\mathbf{w}_1,\mathbf{z}_1)dz_1 dw_1\cdot \mathbf{da}$$

The tensors $\mathbf{T}_\infty$ and $\mathbf{T}_d$ are the background and disturbance stress respectively, $N$ is the number of spheres in each realization of the ensemble of systems, $q_1$ is the single sphere probability function, $\sigma$ is the radius of each sphere, and $\partial\Omega$ is the surface bounding the mixture. The conditional ensemble average, given the position of sphere 1 is denoted $\langle\rangle_1$.

The division of terms between the divergence of the stress $\nabla\cdot\mathbf{T}_f$ and the interaction force $\mathbf{M}_f$ is somewhat arbitrary. We have taken the view that if the divergence of an open quantity appears in the momentum balance, then that quantity belongs in the stress. All other open quantities belong in the interaction force.

**2. A model system.** To calculate the drag in a porous medium, we consider an ensemble of systems like that shown in figure 1. Each system is a closed cylinder filled with a Newtonian fluid, of density $\gamma_f$, in which are suspended $N$ identical rigid spheres. Gravity is neglected. The spheres are fixed in position with respect to some inertial reference frame. The bounding cylinder moves with respect to the spheres along its axis which drives the flow of fluid through the packed bed of spheres. A sufficiently long region of clear fluid is assumed to exist at the ends of the cylinder so that the cylinder may be assumed to move at constant speed $U$. The spheres are assumed to be distributed uniformly through the region they occupy.

The background velocity is in this case $U\mathbf{N}$, with $\mathbf{N}$ the unit vector in the direction of motion of the cylinder. The background pressure is zero. For sufficiently slow motion of the fluid, the disturbance fields satisfy the Stokes equations. When suitably nondimensionalized, these are

$$\nabla \cdot \mathbf{u} = 0 \tag{2.1}$$

$$\nabla \cdot \nabla \mathbf{u} - \nabla p = \mathbf{0} \tag{2.2}$$

with the boundary conditions:

$$\boldsymbol{u} = \mathbf{0} \quad \text{on } \partial\Omega, \qquad \boldsymbol{u} = -\mathbf{N} \quad \text{on } \partial\Omega_i \tag{2.3}$$

where $\partial\Omega_i$ is the surface of the $i^{th}$ sphere.

The solution of this system of equations may be written in terms of the Stokes Green Functions $\mathcal{V}$ and $\mathbf{\Pi}$. The second order tensor $\mathcal{V}$ and the vector $\mathbf{\Pi}$ correspond to the velocity and pressure in the ordinary Stokes equations and are the solutions of the Singular Stokes equations in the fluid region $\Omega_f$. That is, for all $\mathbf{x}$ and $\mathbf{x}_0$ in $\Omega_f$,

$$\nabla \cdot \mathcal{V}(\mathbf{x}, \mathbf{x}_0) = \mathbf{0} \tag{2.4}$$

$$\nabla \cdot \nabla \mathcal{V}(\mathbf{x}, \mathbf{x}_0) - \nabla \mathbf{\Pi}(\mathbf{x}, \mathbf{x}_0) = \delta(\mathbf{x} - \mathbf{x}_0) H_f(\mathbf{x}_0)\mathbf{1} \tag{2.5}$$

For all $\mathbf{x}$ on $\partial\Omega_f$, with $\mathbf{x}_0$ in $\Omega_f$,

$$\mathcal{V}(\mathbf{x}, \mathbf{x}_0) = \mathbf{0} \tag{2.6}$$

In equation 2.5, $\delta(\mathbf{x}) = \delta(x^1)\delta(x^2)\delta(x^3)$ is the three dimensional Dirac delta function and $\mathbf{1}$ is the second order identity tensor.

Given the Stokes Green functions, the solution of the Stokes equations may be written as:

$$\mathbf{u}(\mathbf{x}) = \mathbf{N} \cdot \sum_{i=1}^{N} \int_{|\mathbf{r}|=1} \mathcal{X}^{\mathsf{T}}(\mathbf{x}_i + \mathbf{r}, \mathbf{x}) \cdot \mathbf{r} \, da_r \tag{2.7}$$

$$p(\mathbf{x}) = \mathbf{N} \cdot \sum_{i=1}^{N} \int_{|\mathbf{r}|=1} \nabla_r \mathbf{\Pi}(\mathbf{x}, \mathbf{x}_i + \mathbf{r}) \cdot \mathbf{r} \, da_r \tag{2.8}$$

In these formulae, $\mathcal{X}$ is the third order stress tensor defined as

$$\mathcal{X} = 2\,\mathrm{SYM}\,\nabla\mathcal{V} - \mathbf{1}\mathbf{\Pi} \tag{2.9}$$

**3. Dimensionless drag vector.** We define the drag vector $\mathbf{D}$ as the dimensionless form of the interaction force vector. That is

$$\mathbf{D}(\mathbf{x}) = \frac{\sigma^2}{\mu U}\mathbf{M}_f\,(\mathbf{x}) \tag{3.1}$$

From the formula 1.3, we have for the model system

$$\mathbf{D}(\mathbf{x}) = -n\int_{|\mathbf{y}|=1+} \langle \mathbf{T}_d\rangle_1(\mathbf{x}\,|\mathbf{x}-\mathbf{y}\,)\cdot \mathbf{y}da_y \tag{3.2}$$

The dimensionless disturbance stress $\mathbf{T}_d$ is the stress due to the dimensionless disturbance velocity and pressure. That is

$$\mathbf{T}_d = 2\,\mathrm{SYM}\,\nabla\mathbf{u} - \mathbf{1}p \tag{3.3}$$

Using the Stokes Green function formulations 2.7 and 2.8 of the disturbance fields, we find

$$\mathbf{T}_d(\mathbf{x}) = \sum_{i=1}^{N}\mathbf{N}\cdot\int_{|\mathbf{r}|=1}\mathbf{K}(\mathbf{x},\mathbf{x}_i+\mathbf{r})\cdot\mathbf{r}da_r \tag{3.4}$$

where

$$K^{ijk\ell}(\mathbf{x},\mathbf{y}) = \partial_{x_j}\mathcal{X}^{i\ell k}(\mathbf{y},\mathbf{x}) + \partial_{x_k}\mathcal{X}^{i\ell j}(\mathbf{y},\mathbf{x}) - \delta^{jk}\partial_{y_i}\Pi^{\ell}(\mathbf{x},\mathbf{y}) \tag{3.5}$$

Using these formulae, and the indistinguishability of the spheres, in the equation 3.2 for the drag vector, we find at the origin;

$$\mathbf{D}(0) = -\,\mathbf{N}\cdot n\mathbf{D}_1 - \mathbf{N}\cdot n^2\mathbf{D}_2 \tag{3.6}$$

in which $n$ is the number density. The one sphere conditional integral $\mathbf{D}_1$ is defined as

$$\mathbf{D}_1 = \int_{|\mathbf{y}|=1+}\int_{|\mathbf{r}|=1}\langle\mathbf{K}\rangle_1\,(0,\mathbf{r}-\mathbf{y}\,|-\mathbf{y}\,):\mathbf{ry}da_r da_y \tag{3.7}$$

and the two sphere conditional integral is defined as

$$\mathbf{D}_2 = \int_{\mathbf{x}_2}\int_{|\mathbf{y}|=1+} g\,(-\mathbf{y},\mathbf{x}_2)\int_{|\mathbf{r}|=1}\langle\mathbf{K}\rangle_{12}\,(0,\mathbf{x}_2+\mathbf{r}\,|-\mathbf{y},\mathbf{x}_2\,):\mathbf{ry}da_r da_y dx_2 \tag{3.8}$$

in which $g$ is the pair distribution function and $\langle\mathbf{K}\rangle_{12}$ denotes the conditional ensemble average of $\mathbf{K}$ given the positions of spheres 1 and 2.

**4. Approximation of the one sphere integral.** We may easily approximate the value of the integrand in 3.7 by replacing $\mathbf{K}$ with $\mathbf{K}_1$, the equivalent value for a system consisting of a single sphere in an unbounded fluid. This may be done because the singular solution of the Stokes equations is principally affected by the boundaries nearest the singularity. When the singularity is very near the surface of sphere 1, the value of $\mathbf{K}$ is not affected by the positions of any of the other spheres, unless the surface of the other sphere is also very close to the singularity. Unless the spheres are nearly touching, we may use the following approximation when $\mathbf{r}$ and $\mathbf{y}$ are both unit vectors:

$$\langle \mathbf{K} \rangle_1 (\mathbf{0}, \mathbf{r} - \mathbf{y} \,|{-\mathbf{y}}) \approx \mathbf{K}_1 (\mathbf{0}, \mathbf{r} - \mathbf{y} \,|{-\mathbf{y}}) \tag{4.1}$$

That is, the value of $\langle \mathbf{K} \rangle_1$ at the origin given that sphere 1 is centered at $-\mathbf{y}$ and the singularity is on the surface of sphere 1, is approximately equal to the value calculated by ignoring the presence of any other spheres. Since the solution must be independent of a translation of the coordinates, we also have

$$\mathbf{K}_1 (\mathbf{0}, \mathbf{r} - \mathbf{y} \,|{-\mathbf{y}}) = \mathbf{K}_1 (\mathbf{y}, \mathbf{r} \,|\mathbf{0}) \tag{4.2}$$

The approximation of $\mathbf{D}_1$ is then:

$$\mathbf{D}_1 \approx \int_{|\mathbf{y}|=1^+} \int_{|\mathbf{r}|=1} \mathbf{K}_1 (\mathbf{y}, \mathbf{r} \,|\mathbf{0}) : \mathbf{r}\mathbf{y} \, da_r \, da_y \tag{4.3}$$

Now in terms of the Stokes-Green functions, the dimensionless force $\mathbf{F}$ on a single sphere, due to a uniform flow of unit velocity in the $\mathbf{N}$ direction is

$$\mathbf{F} = \mathbf{N} \cdot \int_{|\mathbf{y}|=1^+} \int_{|\mathbf{r}|=1} \mathbf{K}_1(\mathbf{y}, \mathbf{r} \,|\mathbf{0}) : \mathbf{r}\mathbf{y} \, da_r \, da_y \approx \mathbf{N} \cdot \mathbf{D}_1 \tag{4.4}$$

and from the well known Stokes formula,

$$\mathbf{F} = -\, 6\pi \mathbf{N} \tag{4.5}$$

so we have

$$\mathbf{D}_1 \approx -6\pi \mathbf{1} \tag{4.6}$$

**5. The two sphere conditional integral $\boldsymbol{D_2}$.** To approximate $\mathbf{D}_2$, we note that in the definition 3.8 of $\mathbf{D}_2$, $\langle \mathbf{K} \rangle_{12}$ is evaluated at the origin, given that the origin is a surface point of sphere 1, with the singularity on the surface of sphere 2. In this case, other spheres affect the value of $\langle \mathbf{K} \rangle_{12}$ only if they are near one of these two surface points or if they lie between the origin and the singularity. If we entirely neglect the effect of all spheres except spheres 1 and 2, $\mathbf{D}_2$ may be approximated as

$$\mathbf{D}_2 \approx \int_{|\mathbf{x}_2|>2} \mathbf{F}_2(\mathbf{x}_2)\, dx_2 \tag{5.1}$$

in which

$$\mathbf{F}_2(\mathbf{x}_2) = \int_{|\mathbf{x}|=1^+} \int_{|\mathbf{r}|=1} \mathbf{K}_{12}(\mathbf{x}, \mathbf{x}_2 + \mathbf{r}\,|\mathbf{0}, \mathbf{x}_2) : \mathbf{rx} da_r da_x \tag{5.2}$$

and $\mathbf{K}_{12}$ is $\mathbf{K}$, with only spheres 1 and 2 immersed in an unbounded fluid. The radial distribution function is approximated as

$$g(\mathbf{0}, \mathbf{x}_2) = \begin{cases} 0 & \text{if } |\mathbf{x}_2| < 2 \\ 1 & \text{if } |\mathbf{x}_2| > 2 \end{cases} \tag{5.3}$$

so that the impenetrability of the spheres is accounted for in the simplest way.

When the singularity is near the surface of one of the spheres $\mathbf{K}_{12}$ is accurately approximated by the method of images. The resulting approximation is, for $|\mathbf{x}| = |\mathbf{r}| = 1$,

$$\begin{aligned} &\mathbf{K}_{12}(\mathbf{x}, \mathbf{x}_2 + \mathbf{r}\,|\mathbf{0}, \mathbf{x}_2) \approx \\ &\approx {}^{\mathsf{T}}\left(2\mathbf{xx} \cdot \nabla \mathcal{X}_1^{\mathsf{T}}(\mathbf{x}_2 + \mathbf{r}, \mathbf{x}\,|\mathbf{x}_2) + \mathbf{x}\mathcal{X}_1^{\mathsf{T}}(\mathbf{x}_2 + \mathbf{r}, \mathbf{x}\,|\mathbf{x}_2)\right) + \\ &+ \left(2\mathbf{x} \cdot \nabla \mathcal{X}_1^{\mathsf{T}}(\mathbf{x}_2 + \mathbf{r}, \mathbf{x}\,|\mathbf{x}_2)\mathbf{x} + \mathcal{X}_1^{\mathsf{T}}(\mathbf{x}_2 + \mathbf{r}, \mathbf{x}\,|\mathbf{x}_2)\mathbf{x}\right)^{\mathsf{T}} - \\ &- 2\mathbf{r1r} \cdot \nabla_r \mathbf{\Pi}_1(\mathbf{x}, \mathbf{x}_2 + \mathbf{r}\,|\mathbf{0}) - \mathbf{r1}\mathbf{\Pi}_1(\mathbf{x}, \mathbf{x}_2 + \mathbf{r}\,|\mathbf{0}) \end{aligned} \tag{5.4}$$

where $\mathcal{X}_1$ and $\mathbf{\Pi}_1$ are the stress and pressure respectively due to a point force outside a single sphere immersed in an otherwise unbounded fluid. This problem was solved analytically by Oseen [3].

Using the spherical symmetry to reduce the order of integration, we find

$$\mathbf{N} \cdot \mathbf{D}_2 = \frac{8}{3}\pi^2 \mathbf{N} \int_{\rho=2}^{\infty} \rho^2 \int_{\phi=0}^{\pi} h(\rho, \phi) \sin\phi\, d\phi\, d\rho \tag{5.5}$$

In which the scalar function $h(\rho, \phi)$ is defined as

$$h(\rho, \phi) = \int_{|\mathbf{r}|=1} \mathbf{NN} : \mathbf{K}_{12}(\mathbf{n}(\phi), \rho\mathbf{N} + \mathbf{r}\,|\mathbf{0}, \rho\mathbf{N}) : \mathbf{rn}(\phi)\, da_r \tag{5.6}$$

Using formula 5.6, the integral over all $\rho$ in equation 5.5 does not converge. This is because any spheres that may be located between spheres 1 and 2 are neglected. If these intermediate spheres are taken into account, the magnitude of $h(\rho, \phi)$ is decreased and $\mathbf{N} \cdot \mathbf{D}_2$ will converge.

An additional sphere affects the stress only if it is near the singularity or is interposed between the singularity and the sphere on which the stress is calculated. When $\rho < 4.0$, a third sphere cannot be interposed between spheres 1 and 2. So we divide the region of integration into the near field, $2 \leq \rho \leq 4$ and the field, $\rho > 4$. As a first approximation, we neglect the effect of additional spheres in the near field. In the far field, we assume the magnitude of $h(\rho,\phi)$ is diminished by a weight function $\langle f\rangle_{12}(\rho,n)$:

$$\mathbf{N}\cdot\mathbf{D}_2 \approx$$

$$\approx \tfrac{8}{3}\pi^2\mathbf{N}\Big( \int_{\rho=2}^{4} \rho^2 \int_{\phi=0}^{\pi} h(\rho,\phi)\sin\phi\, d\phi\, d\rho + \\ + \int_{\rho=4}^{\infty} \rho^2 \langle f\rangle_{12}(\rho,n) \int_{\phi=0}^{\pi} h(\rho,\phi)\sin\phi\, d\phi\, d\rho\Big) \tag{5.7}$$

Calculating the first integral numerically, we find

$$\mathbf{D}(0) = \Big(6\pi n + 1.47\times 10^{-3} n^2 + B(n)\Big)\mathbf{N} \tag{5.8}$$

in which

$$B(n) = -\frac{8}{3}\pi^2 n^2 \int_{\rho=4}^{\infty} \rho^2 \langle f\rangle_{12}(\rho,n) \int_{\phi=0}^{\pi} h(\rho,\phi)\sin\phi\, d\phi\, d\rho \tag{5.9}$$

A crude approximation is to simply neglect $B(n)$ on the assumption that interposed spheres will shield the sphere at the origin from the effect of any sphere in the far field. The magnitude of the drag vector was computed with this assumption for solid volume fraction less than 1/2. The result is presented in figure 2 along with experimental data due to Happel and Epstein [4] and Carman [5]. The effective medium approximation calculated by Itoh [6] is also shown for comparison. The approximation fits the data of Happel and Epstein but not that of Carman. It appears then that for solid volume fractions less than about 0.5, setting $B(n)$ to zero is a reasonable approximation. For larger volume fractions, the spheres in the far field have a significant effect which needs to be taken into account.

The reason for this is easily understood. Setting $B(n)$ to zero is equivalent to eliminating all the spheres from each system except a small cloud centered on the origin. If these remaining spheres are packed closely together, the fluid will tend to flow around the cloud rather than through it; thus reducing the drag on the sphere at the center.

Also presented in figure 2 is an approximation for solid volume fraction greater than 0.5 in which we estimate the weight function $\langle f\rangle_{12}(\rho,n)$ as the expected value, given that there are spheres centered at the origin and at $\rho\mathbf{N}$, of a function $f(\rho)$ defined as

$$f(\rho) = \begin{cases} 1 & \text{if no sphere centers are in } V(\rho) \\ 0 & \text{Otherwise} \end{cases} \tag{5.10}$$

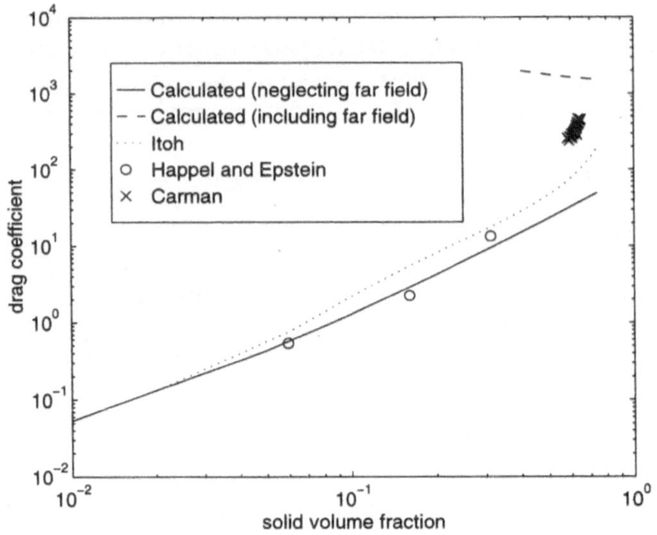


FIG. 2. *Drag Coefficient Including the Far Field for Large Solid Volume Fraction Only.*

$V(\rho)$ is a right circular cylinder of unit radius, on the line segment between the origin and $\rho\mathbf{N}$. That is, in any given system in the ensemble, if the line segment connecting the centers of spheres 1 and 2 intersects any other sphere, sphere 1 is shielded from sphere 2. Otherwise, the effect of sphere 2 is assumed to be the same as if no other spheres were present.

This definition of $f(\rho)$ is equivalent to

$$f(\rho) = \prod_{i=3}^{N} \left( 1 - \int_{\mathbf{y} \in V(\rho)} \delta(\mathbf{y} - \mathbf{x}_i)\, d^3 y \right) \tag{5.11}$$

Using the indistinguishability and impenetrability of the spheres, we find for $\rho \geq 4$ and large $N$,

$$\langle f \rangle_{12}(\rho) \approx \sum_{k=0}^{\infty} \frac{1}{k!} \Big( -\pi n (\rho - 4) \Big)^k = e^{-\pi n (\rho - 4)} \tag{5.12}$$

Using this approximation of the weight function, $B(n)$ becomes

$$B(n) = -\frac{8}{3} \pi^2 n^2 e^{4\pi n} \int_{\rho=4}^{\infty} \rho^2 e^{-\pi n \rho} \int_{\phi=0}^{\pi} h(\rho, \phi) \sin\phi\, d\phi\, d\rho \tag{5.13}$$

The effect of the spheres in the far field is overestimated in this approximation. This indicates that either the weight function calculated from the shadow approximation underestimates the shielding effect of the spheres intervening between the sphere at the origin and the sphere at $\rho\mathbf{N}$, or that the image approximation to the two sphere stress overestimates the magnitude of the stress. Both of these possible deficiencies may be treated by further refinement of the numerical approximation used.

## REFERENCES

[1] M.F. Hurwitz, *Hydrodynamic Interactions of Many Rigid Spheres*, Cornell University, Doctoral Dissertation, 1996.

[2] D.A. Drew, *Mathematical Modeling of Two-Phase Flow*, Ann, Rev. Fluid Mech., 15, 1983, 261–91.

[3] C.W. Oseen, *Neuere Methoden und Ergebnisse in der Hydrodynamik*, Akademische Verlagsgesellschaft M. B. H., 1927.

[4] J. Happel, N. Epstein, *Cubical Assemblages of Uniform Spheres, Industrial and Engineering Chemistry*, 46, 1954, 1187–1194.

[5] P.C. Carman, *Fluid Flow Through Granular Beds*, Trans. Inst. Chem. Eng., 15, 1937, 150–166.

[6] S. Itoh, *The Permeability of a Random Array of Identical Rigid-Spheres*, Journal of the Physical Society of Japan, 52, 1983, 2379–2388.

# ENSEMBLE AVERAGING TECHNIQUES FOR DISPERSE FLOWS

A. PROSPERETTI*

**Abstract.** In a number of recent papers by the author and co-workers, a new formalism for the derivation of averaged equations for disperse multiphase flow was developed. The techniques used in those studies are reviewed here in greater detail than was possible in the original publications. Some examples of the application of the formalism to the dilute case are then shown. Considerations on the numerical implementation of the method and on the incorporation of turbulence in the general framework are also given.

**1. Introduction.** In several earlier papers (Zhang and Prosperetti 1994a, 1994b, 1997; Prosperetti and Zhang 1995, 1996) we have developed a new method of ensemble averaging for disperse multiphase flow. The method is very flexible and can be applied unchanged to a wide variety of situations. Rigorous closure is available for the dilute limit, for both uniform and non-uniform suspensions. Furthermore, a whole series of *ad hoc* approximations can be developed within the framework of the method. Finally, the closure problem presents itself in a form that can be effectively tackled by means of direct numerical simulation.

The mathematical techniques that have proved effective to derive these results have been developed over time. Furthermore, it was deemed unsuitable to include some of their more elementary aspects in research papers, with the consequence that a full comprehension of those studies requires a background that not everybody interested in the results necessarily has. For this reason, we propose to give here a unified account starting from fairly elementary principles.

Our method is based on ensemble averaging but differs from earlier work in several respects. In the first place, we use phase ensemble averaging for the continuous phase (i.e., we average over all the configurations such that at time $t$ the position $\mathbf{x}$ is occupied by the continuous phase), but not for the disperse phase. For the latter we introduce a "particle" ensemble average in which global particle attributes (e.g. the velocity of the center of mass) are averaged directly. A similar idea can be found in Anderson and Jackson (1967), who however used volume averaging. Particle averaging seems to have many advantages as will be discussed below.

Secondly, by explicitly and systematically using the "small-particle approximation" (Zhang and Prosperetti 1994a), i.e. the assumption that the particle size $a$ is small compared with the macroscopic characteristic length $L$, we can considerably simplify the derivation of the averaged equations. Thirdly, unlike some of the earlier work (e.g. Hinch 1977), we derive equa-

* Department of Mechanical Engineering, The Johns Hopkins University, Baltimore, MD 21218.

tions of the so-called "two-fluid" form widely used in Engineering (see e.g. Drew 1983; Wallis 1991).

The method of ensemble averaging is not new in its application to multi-phase flows (see e.g. Batchelor 1972, 1974; Drew 1983). The advantage with respect to time- or volume-averaging is complete generality in the sense that it does not presuppose the separation of scales that the other two approaches require, namely the existence of "a time interval $\Delta t$ ... large enough to smooth out the local variations of properties yet small compared to the macroscopic time constant of the unsteadyness of the bulk fluid" (Ishii 1975, p. 63), or of "an elementary macrovolume $dV$ ... the characteristic linear dimensions of which are many times greater than the nonuniformities $a$ ... but at the same time much less than the characteristic macrodimensions $L$ of the problem" (Nigmatulin 1979). On the negative side, the practical calculation of ensemble averages in situations in which they are not equivalent to simpler ones (e.g. time or volume averages) is more difficult, although the rapidly increasing computational capabilities hold promise to make them directly accessible in the not-too-distant future.

**2. Probabilities.** We consider an ensemble of macroscopically identical suspensions of $N$ spherical particles in a fluid continuous phase. We use the word *configuration* and the symbol $\mathcal{C}^N$ to indicate the set of values of a number of quantities sufficient to specify uniquely the dynamical state of the system at time $t$. In particular, $\mathcal{C}^N$ will include the current values of the particles' degrees of freedom such as position of the center $\{\mathbf{y}^\alpha\}$, $\alpha = 1, 2, ..., N$, center-of-mass velocity $\{\mathbf{w}^\alpha\}$, and possibly others, such as angular velocity, orientation for non-homogeneous particles, or radius for spherical bubbles. For Stokes or inviscid irrotational flow conditions with deterministic boundaries, the continuous-phase degrees of freedom are entirely dependent on those of the particles and need not be included in the specification of $\mathcal{C}^N$. At finite Reynolds numbers, however, they must be explicitly included in $\mathcal{C}^N$ to specify a configuration uniquely. We will return on this point in section 11 and for the time being only include the particle variables in $\mathcal{C}^N$.

The set of all possible values of the variables included in $\mathcal{C}^N$ defines the phase space of the system. The ensemble that we consider consists of a large number of copies of this system (or realizations of the flow) that only differ in the details of the flow, but not in what may be considered its macroscopic features. The evolution in time of each member of the ensemble is represented by the trajectory of a point in phase space. Due to their large number, the members of the ensemble may be considered distributed in phase space according to a probability density function $P(\mathcal{C}^N;t) \equiv P(N;t)$. Specifically,

$$P(N;t)\, d\mathcal{C}^N \equiv P(\mathbf{y}^1, \mathbf{w}^1, \ldots; t)\; d^3y^1\, d^3w^1 \ldots \qquad (2.1)$$

represents the probability of finding, at time $t$, the system in a configuration in which the first particle is within $d^3y^1$ of $\mathbf{y}^1$, etc. This probability density evolves in time according to the well-known equation expressing the conservation of the number of realizations constituting the ensemble:

$$\frac{\partial P}{\partial t} + \sum_{\alpha=1}^{N} [\nabla_\alpha \cdot (\mathbf{w}^\alpha P) + \Delta_\alpha \cdot (\dot{\mathbf{w}}^\alpha P)] = 0 , \tag{2.2}$$

where we have introduced the abbreviations

$$\nabla_\alpha = \nabla_{\mathbf{y}^\alpha} , \qquad \Delta_\alpha = \nabla_{\mathbf{w}^\alpha} . \tag{2.3}$$

We have written (2.2) for the case in which a configuration is defined in terms of positions and velocities. If additional variables are needed, the corresponding differential terms must be added. Conversely, for Stokes flow in which the particle velocity in most cases depends on the position only, the velocity derivative terms should be dropped.

The probability distribution defined in (2.1) treats the particles as distinguishable in the sense that, for example, the third set of arguments of $P$ gives the dynamical state of the particle carrying the label 3. In the system we are considering, however, the particles are identical. This implies that, if we solve Eq. (2.2) starting from initial states that only differ by a permutation of the particles' initial condition, at any later time $t$ we will find states with the corresponding permutation, and such that the numerical values of the two $P$'s are identical. Thus even though, by looking at the two $P$'s, we are able to tell which particle is where, we should not consider the two states as different from each other. There are several ways to deal with this feature. One that is particularly convenient is to retain the distinguished-particle probability $P$, and to introduce suitable corrections to the normalization relations to account for the indistinguishability of the states. The way to achieve this objective is best explained by considering the simple situation of two particles constrained to the interval [0,1]. Let $P(a, b)$ be the probability density for finding particle 1 between $a$ and $a + da$, and particle 2 between $b$ and $b + db$. The situation in which the position of the two particles is interchanged is a possible one, but it is not a different state of the system, which implies in particular that $P(b, a)$ is numerically equal to $P(a, b)$. Since each particle can be anywhere in the unit interval, the proper normalization must involve the integration of $P$ with both particle positions ranging between 0 and 1. However, in so doing, each physically distinct state is counted twice, once with particle 1 to the left of 2, and the other one with the particles interchanged. We can avoid this double counting by dividing the integral of $P$ by the degeneracy of each state, namely by imposing that

$$\frac{1}{2!} \int_0^1 db \int_0^1 da \, P(a, b) = 1 . \tag{2.4}$$

The argument can be generalized to $N$ particles for which

$$\frac{1}{N!}\int dC^N\, P(N;t) = 1 \quad , \tag{2.5}$$

with the integration over the complete range for each variable. This technique enables us to make use of a probability distribution in which the particles are distinguished, without the need to formally define and explicitly deal with a symmetrized one in which they are not.

In addition to $P(N;t)$ we shall make use of the reduced $K$-particle probability distribution in which the configuration of the first $K$ particles is prescribed:

$$P(K;t) = \frac{1}{(N-K)!}\int dC^{N-K}\, P(N;t) \ . \tag{2.6}$$

Here the integration is over the configuration of the last $N-K$ particles and the factor $(N-K)!$ accounts for the indistinguishability of states that only differ by a permutation of these $N-K$ particles. From (2.5), it is readily shown that

$$\int dC^K\, P(K;t) = \frac{N!}{(N-K)!} \quad , \tag{2.7}$$

the meaning of which is apparent when it is realized that $N!/(N-K)!$ is the product of the number of ways in which a group of $K$ particles can be chosen out of $N$, and the number of possible permutations of each $K$-particle group. We shall be especially concerned with the one-particle probability density function $P(1_y;t) \equiv P(\mathbf{y},\mathbf{w},t)$ which, according to (2.7), satisfies

$$\int d^3y \int d^3w\, P(\mathbf{y},\mathbf{w},t) = N \ . \tag{2.8}$$

This relation justifies the following definition of particle number density $n$:

$$n(\mathbf{y},t) = \int d^3w\ P(\mathbf{y},\mathbf{w},t) \ . \tag{2.9}$$

The conditional probability $P(N-K|K;t)$ of having the last $N-K$ particles in a certain configuration, given that the first $K$ have another configuration, is defined by

$$P(N-K|K;t) = \frac{P(N;t)}{P(K;t)} , \tag{2.10}$$

and, as a consequence, satisfies the normalization condition

$$\int dC^{N-K} P(N-K|K;t) = (N-K)! \ , \tag{2.11}$$

as could be expected.

The evolution equation for $P(K;t)$ can be obtained by averaging that for $P(N;t)$, Eq. (2.2), over the configurations of the last $N-K$ particles. The result is

$$\frac{\partial P(K;t)}{\partial t} + \sum_{\alpha=1}^{K}\{\nabla_\alpha \cdot [\mathbf{w}^\alpha P(K;t)] + \Delta_\alpha \cdot [\ll \dot{\mathbf{w}}^\alpha \gg_K P(K;t)]\} = 0, \tag{2.12}$$

where, for $\alpha = 1, 2, \ldots, K$,

$$\ll \dot{\mathbf{w}}^\alpha \gg_K = \frac{1}{(N-K)!}\int P(N-K|K;t)\, \dot{\mathbf{w}}^\alpha \, dC^{N-K}. \tag{2.13}$$

As a simple application of this result we can derive the evolution equation for the number density $n$. By taking the time derivative of (2.9) and using (2.12) with $K=1$ to eliminate $\partial P(1;t)/\partial t$, we have

$$\frac{\partial n}{\partial t} = -\int d^3w \, \{\nabla \cdot [\mathbf{w} P(1;t)] + \Delta \cdot [\ll \dot{\mathbf{w}} \gg_1 P(1;t)]\}. \tag{2.14}$$

The last term is an exact differential and integrates to zero since $P$ is assumed to vanish at infinity in velocity space. If we define the mean center-of-mass velocity $\overline{\mathbf{w}}$ by

$$\overline{\mathbf{w}}(\mathbf{y},t) = \frac{1}{n(\mathbf{y},t)}\int d^3w \, \mathbf{w}\, P(\mathbf{y},\mathbf{w};t), \tag{2.15}$$

we thus find

$$\frac{\partial n}{\partial t} + \nabla \cdot (n\, \overline{\mathbf{w}}) = 0. \tag{2.16}$$

From the evolution equations (2.2) for $P(N;t)$ and (2.12) for $P(K;t)$ one can derive the evolution equation for the conditional probability $P(N-K|K;t)$:

$$\begin{aligned} \frac{\partial P(N-K|K;t)}{\partial t} &+ \sum_{\alpha=1}^{N}\Big\{\nabla_\alpha \cdot [\mathbf{w}^\alpha P(N-K|K;t)] \\ &\qquad + \Delta_\alpha \cdot [\dot{\mathbf{w}}^\alpha P(N-K|K;t)]\Big\} \\ &= \frac{P(N-K|K;t)}{P(K;t)} \sum_{\alpha=1}^{K}\Big\{\Delta_\alpha \cdot [\ll \dot{\mathbf{w}}^\alpha \gg_K P(K;t)] \\ &\qquad - \dot{\mathbf{w}}^\alpha \cdot \Delta_\alpha P(K;t)\Big\}. \end{aligned} \tag{2.17}$$

This relation is necessary to derive the evolution equations for the conditional averages defined below.

**3. Volume fractions.** As shown explicitly e.g. by Celmins (1988), the volume fraction defined on the basis of volume averaging is a highly oscillatory quantity unless the averaging volume is several inter-particle distances in size. In the context of ensemble averaging, these spurious oscillations are removed by introducing the very natural definition of volume fraction as the probability that the point $\mathbf{x}$ is occupied by the phase of interest at the time $t$.

To give a formal definition of this concept we introduce the characteristic, or indicator, functions $\chi_{C,D}(\mathbf{x};N)$ for the continuous and disperse phases respectively (see e.g. Lundgren 1972; Drew 1983). For example, the function $\chi_C$ equals 1 at the point $\mathbf{x}$ when, given the configuration $\mathcal{C}^N$, $\mathbf{x}$ is in the continuous phase, and 0 otherwise. Note that $\chi_{C,D}$ do not depend on time directly, but only on the geometry of the instantaneous $\mathcal{C}^N$. Of course, as the configuration evolves, the value of the characteristic function at a given point will also change in general. The volume fractions $\beta_{C,D}$ of the phases are then defined as averages of the characteristic functions over the probability $P(N;t)$:

$$\beta_{C,D}(\mathbf{x},t) = \frac{1}{N!}\int d\mathcal{C}^N P(N;t)\,\chi_{C,D}(\mathbf{x};N)\,. \tag{3.1}$$

If the interfaces between the continuous and disperse phases have zero measure, as we assume, evidently one has $\chi_C + \chi_D = 1$ for all points in space, from which it follows that

$$\beta_C + \beta_D = 1\,. \tag{3.2}$$

For a suspension of equal spherical particles of radius $a$, $\chi_D$ can be explicitly represented as (Lundgren 1972):

$$\chi_D(\mathbf{x};N) = 1 - \chi_C(\mathbf{x},N) = \sum_{\alpha=1}^{N} H\left(a - |\mathbf{x}-\mathbf{y}^\alpha|\right), \tag{3.3}$$

where $H$ is the Heaviside distribution. Similar relations can be written for other particle shapes. For future reference we note the result

$$\nabla\chi_C = \sum_{\alpha=1}^{N} \delta(a - |\mathbf{x}-\mathbf{y}^\alpha|)\,\frac{\mathbf{x}-\mathbf{y}^\alpha}{a}\,. \tag{3.4}$$

By using the representation (3.3) one can readily relate $\beta_D$ to the particle number density $n$ (Lundgren 1972) defined in (2.9):

$$\begin{aligned}\beta_D(\mathbf{x},t) &= \frac{1}{N!}\sum_{\alpha=1}^{N}\int d\mathcal{C}^N H(a-|\mathbf{x}-\mathbf{y}^\alpha|)P(N;t)\\ &= \frac{1}{(N-1)!}\int d\mathcal{C}^1 H(a-|\mathbf{x}-\mathbf{y}^{(1)}|)\int d\mathcal{C}^{N-1}P(N;t)\\ &= \int_{|\mathbf{x}-\mathbf{y}|\le a} d^3y \int d^3w\, P(\mathbf{y},\mathbf{w};t) = \int_{|\mathbf{x}-\mathbf{y}|\le a} d^3y\, n(\mathbf{y},t)\,,\end{aligned} \tag{3.5}$$

where $d\mathcal{C}^1 \equiv d^3y^1 d^3w^1$ and the superscript 1 has been dropped in the last line. The first step is justified by the fact that the particles are all identical so that each one of them gives the same contribution to the sum. The second step follows from the definition (2.6) of reduced probability distribution. Upon expanding $n(\mathbf{y},t)$ in a Taylor series around $\mathbf{x}$ we find

$$\beta_D = nv + \frac{1}{10} va^2 \nabla^2 n + o\left(\frac{a^2}{L^2} nv\right), \tag{3.6}$$

where $L$ is the scale of variation of the averaged quantities and $v = \frac{4}{3}\pi a^3$ is the particle volume. This result shows the relation between the true volume fraction and the often-used approximation $\beta_D \simeq nv$.

Averages of the characteristic functions over the conditional probabilities $P(N-K|K;t)$ give the conditional volume fractions:

$$\beta^{(K)}_{C,D}(\mathbf{x},t|K) = \frac{1}{(N-K)!} \int d\mathcal{C}^{N-K} P(N-K|K;t)\, \chi_{C,D}(\mathbf{x};N). \tag{3.7}$$

For example, $\beta^{(K)}_D(\mathbf{x},t|K)$ gives the probability that the point $\mathbf{x}$ at time $t$ is in the disperse phase when the first $K$ particles are in the configuration $\mathcal{C}^K$. In particular we find

$$\beta^{(1)}_D(\mathbf{x},t|\mathbf{y},\mathbf{w}) = H(a-|\mathbf{x}-\mathbf{y}|) + \int_{|\mathbf{x}-\mathbf{y}'|\le a} d^3y' \int d^3w'\, P(\mathbf{y}'\mathbf{w}'|\mathbf{y},\mathbf{w};t), \tag{3.8}$$

where $P(\mathbf{y}'\mathbf{w}'|\mathbf{y},\mathbf{w};t) = P(\mathbf{y},\mathbf{w},\mathbf{y}',\mathbf{w}';t)/P(\mathbf{y},\mathbf{w};t)$ is the two-particle conditional probability.

**4. Particle averaging.** One of the distinctive features of the present theory is the use of two different averaging procedures, particle and phase averaging. The former is applicable to quantities that pertain to each particle as a whole. Examples are center-of-mass velocity and acceleration, angular velocity, and others.

Let $g^\alpha(N;t)$ be any such quantity, where the notation implies that the value of $g$ for the particle $\alpha$ depends, in general, on the entire configuration. Then we define the average value $\overline{g}(\mathbf{x},t)$ as the mean value of the quantity $g$ whenever any particle is centered at $\mathbf{x}$ at time $t$ irrespective of the velocity of that particle and of the configuration of all the other $N-1$ particles:

$$n\,\overline{g}(\mathbf{x},t) = \frac{1}{N!} \int d\mathcal{C}^N P(N;t) \left[\sum_{\alpha=1}^{N} \delta(\mathbf{x}-\mathbf{y}^\alpha)\, g^\alpha(N;t)\right]. \tag{4.1}$$

Because of the identity of the particles, from this definition we find

$$\begin{aligned} \overline{g}(\mathbf{x},t) &= \frac{1}{n(\mathbf{x},t)} \frac{1}{(N-1)!} \\ &\times \int d^3w^1 \int d\mathcal{C}^{N-1} P(\mathbf{x},N-1;t)\, g^1(\mathbf{x},N-1;t), \end{aligned} \tag{4.2}$$

where the notation implies that the center of particle 1 is at $\mathbf{x}$. If the quantity $g^\alpha$ does not depend explicitly on the configuration of the other particles, this definition reduces to

$$\overline{g}(\mathbf{x},t) = \frac{1}{n(\mathbf{x},t)} \int d^3w \, P(\mathbf{x},\mathbf{w};t)\, g^1(\mathbf{x},\mathbf{w},t). \tag{4.3}$$

The center-of-mass velocity $\overline{\mathbf{w}}$ previously introduced in (2.15) is clearly a special case of this relation.

In general, any one of the quantities $g^\alpha(N;t)$ depends on time both directly and indirectly through the evolution of the configuration. Denoting by $\hat{\partial}/\hat{\partial}t$ the derivative with respect to the explicit time dependence (i.e. keeping the configuration fixed), and by a dot the total time derivative, we therefore have

$$\dot{g}^\alpha = \frac{\hat{\partial} g^\alpha}{\hat{\partial} t} + \sum_{\beta=1}^{N} \left( \mathbf{w}^\beta \cdot \nabla_\beta \, g^\alpha + \dot{\mathbf{w}}^\beta \cdot \Delta_\beta \, g^\alpha \right) . \tag{4.4}$$

Upon differentiating with respect to time the quantity $n\,\overline{g}$ according to the definition (4.2), using this relation, and the evolution equation (2.2) for the probability $P$, we find

$$\begin{aligned} \frac{\partial}{\partial t}(n\,\overline{g}) &= n\, \overline{\frac{\hat{\partial} g^1}{\hat{\partial} t}} - \frac{1}{(N-1)!} \\ &\times \int d^3w^1 \int d\mathcal{C}^{N-1} g^1 \sum_{\beta=1}^{N} \left[ \nabla_\beta \cdot (\mathbf{w}^\beta P) + \Delta_\beta \cdot (\dot{\mathbf{w}}^\beta P) \right]. \end{aligned} \tag{4.5}$$

The terms corresponding to $\beta = 2, 3, \ldots, N$ can be integrated by parts and one is left with

$$\frac{\partial}{\partial t}(n\,\overline{g}) + \nabla \cdot (n\,\overline{\mathbf{w}g}) = n\,\overline{\dot{g}^1}\,, \tag{4.6}$$

with $\dot{g}^1$ given by (4.4). This result is formally identical to one familiar from the kinetic theory of gases.

In extensions of the analysis beyond the first order in $\beta_D$, conditional particle averages are required (Zhang 1997). In order to simplify the equations we assume a Stokes flow situation, in which the particle velocities are dependent variables and, as an example, we consider the field $g^\alpha$ at $\mathbf{x}$ averaged subject to the presence of a particle at $\mathbf{z}$. The definition (4.1) is readily generalized to this case by extending the summation to products of delta functions with poles at $\mathbf{x}$ and $\mathbf{z}$. As many terms should be included in the summation as there are ordered groups of two objects in a set of $N$. The number of ways in which this can be arranged is the number of ordered particle pairs extracted from $N$ particles, $S$ say, multiplied by the permutations of the remaining $N-2$. Thus we write

$$
\begin{aligned}
n^{(1)} \overline{g}^{(1)}(\mathbf{x}, t|\mathbf{z}) &= \frac{1}{S(N-2)!} \\
&\times \int dC^N \sum_{\alpha=1}^{N} \sum_{\beta \neq \alpha} \frac{P(N;t)}{P(\mathbf{y}^\beta;t)} \delta(\mathbf{x}-\mathbf{y}^\alpha)\, \delta(\mathbf{z}-\mathbf{y}^\beta)\, g^\alpha(N;t),
\end{aligned} \tag{4.7}
$$

where $n^{(1)} = n^{(1)}(\mathbf{x}, t|\mathbf{z})$, to be defined below, is the conditional number density. The introduction of the one-particle probability in the right-hand side is a matter of definition. Its motivation here is that, by definition, $P(N;t)/P(\mathbf{y}^\beta;t) = P(N-1;t|\mathbf{y}^\beta)$ is the probability conditional with one particle being at $\mathbf{y}^\beta$, which is the quantity of interest.

Since each one of the $S$ terms of the sum gives the same contribution, we can simply consider $\mathbf{x}$ as the position of particle 1, $\mathbf{z}$ that of particle 2, and multiply by $S$ to find

$$
n^{(1)} \overline{g}^{(1)}(\mathbf{x}, t|\mathbf{z}) = \frac{1}{(N-2)!} \int dC^{N-2}\, P(\mathbf{x}, N-2; t|\mathbf{z})\, g^1(\mathbf{x}, \mathbf{z}, N-2; t). \tag{4.8}
$$

The conditional particle number density $n^{(1)}$ is determined by setting $g^1 = 1$ and is therefore

$$
\begin{aligned}
n^{(1)}(\mathbf{x}, t|\mathbf{z}) &= \frac{1}{(N-2)!} \int dC^{N-2}\, P(\mathbf{x}, N-2; t|\mathbf{z}) \\
&= \frac{1}{P(\mathbf{z};t)} \frac{1}{(N-2)!} \int dC^{N-2}\, P(N;t) \\
&= \frac{P(\mathbf{x}, \mathbf{z}; t)}{P(\mathbf{z};t)} = P(\mathbf{x}, t|\mathbf{z}),
\end{aligned} \tag{4.9}
$$

i.e. the one-particle conditional probability. With velocity as an independent variable, one can define other conditional averages and the proper definition of $n^{(1)}$ might then involve integrals of the one-particle conditional probability over one or both velocity variables.

The introduction of the particle average (4.2) has been motivated here for quantities that pertain to the particle as a whole. As will be argued at the end of section 5, the applicability and usefulness of this type of averaging extends considerably beyond these simplest examples of such quantities.

**5. Phase averaging.** For a field quantity $f_{C,D}(\mathbf{x}, t; N)$ pertaining to the continuous or disperse phase, the phase averages are defined by averaging over all the configurations in which the point $\mathbf{x}$ is in the proper phase:

$$
< f_{C,D} > (\mathbf{x}, t) = \frac{1}{N! \beta_{C,D}} \int dC^N\, f_{C,D}(\mathbf{x}, t; N)\, \chi_{C,D}(\mathbf{x}; N)\, P(N;t). \tag{5.1}
$$

This form of averaging is attractive because it can be used also when the field $f$ is defined in only one of the phases. An example would be the pressure in a suspension of rigid particles for which $p_C$ has the usual meaning while $p_D$ would require a special interpretation. Furthermore, the use of separate averages for the two phases leads naturally to a two-fluid formulation for the averaged equations. A consequence of this definition, however, is that differentiation and averaging do not commute, as will be shown presently.

The relations needed for the two phases are somewhat different and, accordingly, we give separate treatments.

**5.1. Continuous phase.** From the definition (5.1), we can calculate the gradient of the continuous-phase average field $< f_C >$:

$$\begin{aligned}
&\nabla(\beta_C < f_C >) \\
&= \frac{1}{N!}\int d\mathcal{C}^N \, (\chi_C \nabla f_C + f_C \nabla \chi_C)\, P(N;t) \\
&= \beta_C < \nabla f_C > + \frac{1}{N!}\int d\mathcal{C}^N \, f_C \sum_{\alpha=1}^{N} \delta(a - |\mathbf{x}-\mathbf{y}^\alpha|)\, \frac{\mathbf{x}-\mathbf{y}^\alpha}{a}\, P(N;t) \\
&= \beta_C < \nabla f_C > + \frac{1}{(N-1)!}\int d\mathcal{C}^N \, f_C \, \delta(a - |\mathbf{x}-\mathbf{y}^1|)\, \frac{\mathbf{x}-\mathbf{y}^1}{a}\, P(N;t) \\
&= \beta_C < \nabla f_C > + \frac{1}{(N-1)!}\int d\mathcal{C}^1 \, \delta(a - |\mathbf{x}-\mathbf{y}^1|)\, \frac{\mathbf{x}-\mathbf{y}^1}{a}\, P(1;t) \\
&\qquad \times \int d\mathcal{C}^{N-1} \, f_C \, P(N-1|1;t)\,.
\end{aligned} \tag{5.2}$$

The first step follows directly from the definition (5.1), the second one from the expression (3.4) for $\nabla\chi_C$, the third one from the identity of the particles as in (3.5), and the last one from the definition of conditional probability. The presence of $f_C$ indicates that, in evaluating the delta function, the point $\mathbf{x}$ must be taken on the side of the sphere's surface exposed to the continuous phase. The last expression can be written in a more compact form by introducing the conditional averages:

$$\begin{aligned}
< f_{C,D} >_K (\mathbf{x}, t|K) &= \frac{1}{(N-K)!\,\beta_{C,D}^{(K)}} \\
&\times \int d\mathcal{C}^{N-K} \, f_{C,D}(\mathbf{x},t;N)\, \chi_{C,D}(\mathbf{x};N)\, P(N-K|K;t)\,.
\end{aligned} \tag{5.3}$$

It is an immediate consequence of the definitions that

$$\beta_{C,D} < f_{C,D} > = \frac{(N-K)!}{N!}\int d\mathcal{C}^K \, \beta_{C,D}^{(K)} < f_C >_K \, P(K;t)\,. \tag{5.4}$$

By using the definition (5.3), and noting that the point $\mathbf{x}$ is on the external surface of the sphere so that $\beta_C^{(1)} = 1$ and $\chi_C = 1$, (5.2) may be more compactly rewritten as

$$\begin{aligned} \nabla(\beta_C < f_C >) = \beta_C < \nabla f_C > + \int_{|\mathbf{x}-\mathbf{y}|=a} dS_y \, \mathbf{n}_y \\ \times \int d^3w < f_C >_1 (\mathbf{x}, t|\mathbf{y}, \mathbf{w}) \, P(\mathbf{y}, \mathbf{w}; t) , \end{aligned} \tag{5.5}$$

where $\mathbf{n}_y$ is the unit normal directed out of the particle. It should be noted that the integral in this relation is over all the particles that touch the point $\mathbf{x}$.

With $f_C = 1$, from this relation one finds the following expression for $\nabla\beta_C$:

$$\nabla\beta_C = -\nabla\beta_D = \int_{|\mathbf{x}-\mathbf{y}|=a} dS_y \, \mathbf{n}_y \, n(\mathbf{y}, t) . \tag{5.6}$$

Upon expanding the derivative in the left-hand side of (5.5) and using this result to express $\nabla\beta_C$, we may also write

$$\begin{aligned} \nabla < f_C > = < \nabla f_C > + \frac{1}{\beta_C} \int_{|\mathbf{x}-\mathbf{y}|=a} dS_y \, \mathbf{n}_y \\ \times \int d^3w \; P(\mathbf{y}, \mathbf{w}; t) \, [< f_C >_1 (\mathbf{x}, t|\mathbf{y}, \mathbf{w}) - < f_C > (\mathbf{x}, t)] . \end{aligned} \tag{5.7}$$

A similar approach can be followed for the time derivative of the phase average $< f_C >$. Starting from the definition (5.1) we find

$$\begin{aligned} \frac{\partial}{\partial t}(\beta_C < f_C >) &= \frac{1}{N!} \int d\mathcal{C}^N \left[ \frac{\hat{\partial} f_C}{\hat{\partial} t} P(N;t) + f_C \frac{\partial P(N;t)}{\partial t} \right] \chi_C(\mathbf{x}; N) \\ &= \beta_C \left\langle \frac{\hat{\partial} f_C}{\hat{\partial} t} + \sum_{\alpha=1}^{N} (\mathbf{w}^\alpha \cdot \nabla_\alpha f_C + \dot{\mathbf{w}}^\alpha \cdot \Delta_\alpha f_C) \right\rangle \\ &\quad + \frac{1}{N!} \int d\mathcal{C}^N f_C \, P(N;t) \sum_{\alpha=1}^{N} \mathbf{w}^\alpha \cdot \nabla_\alpha \chi_C . \end{aligned} \tag{5.8}$$

In the first line, the time derivative is only with respect to the explicit time dependence of $f_C$ as the particle degrees of freedom are integration variables. The next step follows from the evolution equation (2.2) of $P(N;t)$ and an integration by parts in phase space. The first group of terms in the right-hand side is clearly the partial derivative of $f_C$ with respect to time keeping $\mathbf{x}$, but not the configuration $\mathcal{C}^N$, fixed:

$$\frac{\partial f_C}{\partial t} = \frac{\hat{\partial} f_C}{\hat{\partial} t} + \sum_{\alpha=1}^{N} (\mathbf{w}^\alpha \cdot \nabla_\alpha f_C + \dot{\mathbf{w}}^\alpha \cdot \Delta_\alpha f_C) . \tag{5.9}$$

The notation $\partial f_C/\partial t$ for this quantity is analogous to (4.4). By introducing also the representation (3.4) of $\nabla\chi_C$ we have

$$\frac{\partial}{\partial t}(\beta_C < f_C >) = \beta_C < \frac{\partial f_C}{\partial t} > - \int_{|\mathbf{y}-\mathbf{x}|=a} dS_y \int d^3w\, P(1_y;t)\, \mathbf{n}_y \cdot \mathbf{w} < f_C >_1 . \tag{5.10}$$

With $f_C = 1$, this relation gives

$$\frac{\partial \beta_C}{\partial t} = - \int_{|\mathbf{y}-\mathbf{x}|=a} dS_y \int d^3w\, P(1;t)\, \mathbf{n}\cdot\mathbf{w}, \tag{5.11}$$

using which (5.10) can equivalently be written

$$\frac{\partial < f_C >}{\partial t} = < \frac{\partial f_C}{\partial t} > - \frac{1}{\beta_C}\int_{|\mathbf{y}-\mathbf{x}|=a} dS_y \int d^3w\, P(1_y;t)\, \mathbf{n}_y \cdot \mathbf{w}\,(< f_C >_1 - < f_C >). \tag{5.12}$$

At this point we can prove a transport theorem that is very useful in the derivation of the averaged equations. Let $\mathbf{u}_C$ be the velocity field of the continuous phase. Then, by using (5.5) and (5.10), we find

$$\frac{\partial}{\partial t}(\beta_C < f_C >) + \nabla\cdot(\beta_C < f_C \mathbf{u}_C >) = \beta_C < \frac{\partial f_C}{\partial t} + \nabla\cdot(f_C\mathbf{u}_C) > - \int_{|\mathbf{x}-\mathbf{y}|=a} dS_y \int d^3w P(1_y;t)\, \mathbf{n}_y \cdot [\mathbf{w} < f_C >_1 - < f_C \mathbf{u}_C >_1] . \tag{5.13}$$

The surface integral is over all the particles the surface of which touches $\mathbf{x}$. For all such particles the velocity field satisfies the kinematic boundary condition

$$\mathbf{n}\cdot\mathbf{w} = \mathbf{n}\cdot\mathbf{u}_C(\mathbf{x}, t; N), \tag{5.14}$$

and, since $\mathbf{w} < f_C >_1 = < \mathbf{w} f_C >_1$, the integral term in (5.13) vanishes identically so that

$$\frac{\partial}{\partial t}(\beta_C < f_C >) + \nabla\cdot(\beta_C < f_C\mathbf{u}_C >) = \beta_C < \frac{\partial f_C}{\partial t} + \nabla\cdot(f_C\mathbf{u}_C) > . \tag{5.15}$$

A formally identical relation holds for the disperse phase:

$$\frac{\partial \beta_D < f_D >}{\partial t} + \nabla\cdot(\beta_D < f_D\mathbf{u}_D >) = \beta_D < \frac{\partial f_D}{\partial t} + \nabla\cdot(f_D\mathbf{u}_D) > . \tag{5.16}$$

A special case, obtained for $f_C = 1$, gives the evolution equation for $\beta_C$:

$$\frac{\partial \beta_C}{\partial t} + \nabla \cdot (\beta_C < \mathbf{u}_C >) = \beta_C < \nabla \cdot \mathbf{u}_C > , \tag{5.17}$$

with a similar relation for the disperse phase.

In addition to the differential-transport operator in the right-hand side of (5.15), the equations to be averaged also contain spatial derivatives, e.g. the divergence of the stress. It is therefore useful to try to simplify the previous result (5.7) that, as it stands, is inconvenient to use as the integration is over all the configurations in which there is a particle touching the point $\mathbf{x}$ rather than over the surface of a fixed particle. Let $\mathbf{r} = \mathbf{x} - \mathbf{y}$ and define

$$F(\mathbf{r}, \mathbf{y}) \equiv P(\mathbf{y}, \mathbf{w}; t) \left[ < f_C >_1 (\mathbf{y} + \mathbf{r}, t | \mathbf{y}, \mathbf{w}) - < f_C > (\mathbf{y} + \mathbf{r}, t) \right] . \tag{5.18}$$

In the integral appearing in (5.7) $|\mathbf{r}| = a$. In many situations of interest $F$ may be expected to depend only weakly on the position $\mathbf{y}$ of the particle center so that one may use Taylor's theorem centered around $\mathbf{x}$ and write

$$F(\mathbf{r}, \mathbf{y}) = F(\mathbf{r}, \mathbf{x}) - \mathbf{r} \cdot \nabla F(\mathbf{r}, \mathbf{x} - \mathbf{h}) , \tag{5.19}$$

where $|\mathbf{h}| < a$ and the Lagrange form of the remainder has been used. In this relation, $\mathbf{r}$ must be interpreted as the distance from the particle center that has been moved from the position $\mathbf{y}$ to $\mathbf{x}$. The idea underlying (5.19) can be further clarified with reference to Fig. 1. The quantity $F$ defined in (5.18), as it appears in the integral (5.7), should be evaluated at the point $P$. Equation (5.19) expresses it as the value at the neighboring point $Q$, plus a correction.

Upon substituting (5.19) into (5.7), we then find

$$\beta_C \nabla < f_C > = \beta_C < \nabla f_C > + \beta_D \mathcal{A}[f_C] - \nabla \cdot (\beta_D \mathcal{L}[f_C]) , \tag{5.20}$$

where

$$\begin{aligned} \beta_D \mathcal{A}[f_C] = & \int d^3 w \, P(\mathbf{x}, \mathbf{w}; t) \\ & \times \int_{|\mathbf{x}-\mathbf{z}|=a} dS_z [< f_C >_1 (\mathbf{z}, t | \mathbf{x}, \mathbf{w}) - \; < f_C > (\mathbf{z}, t)] \, \mathbf{n} , \end{aligned} \tag{5.21}$$

$$\begin{aligned} \beta_D \mathcal{L}[f_C] = a & \int d^3 w \int_{|\mathbf{x}-\mathbf{z}|=a} dS_z \, P(\mathbf{x} - \mathbf{h}, \mathbf{w}; t) \\ & \times \left[ < f_C >_1 (\mathbf{z} - \mathbf{h}, t | \mathbf{x} - \mathbf{h}, \mathbf{w}) - < f_C > (\mathbf{z} - \mathbf{h}, t) \right] \mathbf{n}\,\mathbf{n} , \end{aligned} \tag{5.22}$$

with $\mathbf{h}$ a function of $\mathbf{z}$. Note that, unlike (5.7), the integral in the definition (5.21) of $\mathcal{A}[f_C]$ is taken over the surface of a sphere centered at $\mathbf{x}$.

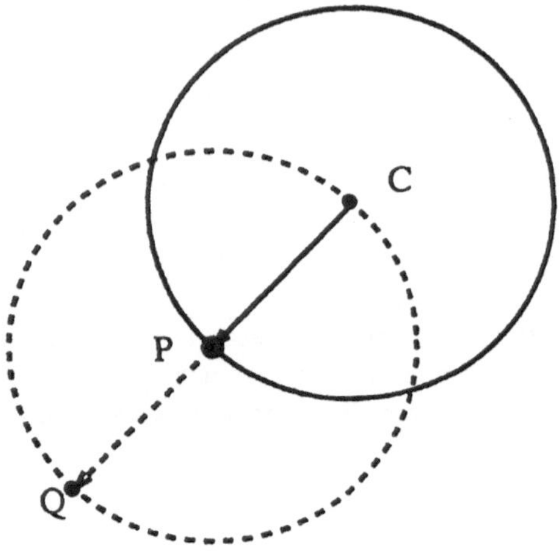


FIG. 1. *Illustration of the approximation used in deriving (5.20). According to (5.7), the value of the integrand at the point P for all the particles touching that point is needed. The solid line shows one such particle centered at C. The contribution of this particle is approximated by the integrand at Q, plus a few terms of the Taylor series expansion. In other words, the contribution of the particle centered at C is approximated by that centered at P.*

Further, both $\mathcal{A}[f_C]$ and $\mathcal{L}[f_C]$ are non-zero only insofar as the flow in the neighborhood of a particle differs from the mean flow. It is therefore clear that these terms will account for the microphysics of the suspension.

For particles small with respect to the scale $L$ of macroscopic variation of the flow, $\mathcal{L}[f_C]$ can be expanded retaining only the first few terms to find (cf. Zhang and Prosperetti 1994a)

$$\beta_D\mathcal{L}[f_C] = \beta_D\mathcal{T}[f_C] + \nabla\cdot(\beta_D\mathcal{S}[f_C]) + \nabla\nabla : (\beta_D\mathcal{R}[f_C]) + O\left(\frac{a^3}{L^3}\beta_D < f_C >_1\right), \tag{5.23}$$

where

$$\beta_D\mathcal{T}[f_C] = a\int d^3w\, P(\mathbf{x},\mathbf{w};t) \times \int_{|\mathbf{x}-\mathbf{z}|=a} dS_z\,[< f_C >_1 (\mathbf{z},t|\mathbf{x},\mathbf{w}) - < f_C > (\mathbf{z},t)]\,\mathbf{n}\,\mathbf{n}, \tag{5.24}$$

$$\beta_D\mathcal{S}[f_C] = -\frac{1}{2}a^2\int d^3w\, P(\mathbf{x},\mathbf{w};t) \times \int_{|\mathbf{x}-\mathbf{z}|=a} dS_z\,[< f_C >_1 (\mathbf{z},t|\mathbf{x},\mathbf{w}) - < f_C > (\mathbf{z},t)]\,\mathbf{n}\,\mathbf{n}\,\mathbf{n}, \tag{5.25}$$

$$\beta_D \mathcal{R}[f_C] = \frac{1}{6} a^3 \int d^3w \, P(\mathbf{x}, \mathbf{w}; t) \times \int_{|\mathbf{x}-\mathbf{z}|=a} dS_z \, [< f_C >_1 (\mathbf{z}, t|\mathbf{x}, \mathbf{w}) - < f_C > (\mathbf{z}, t)] \, \mathbf{nnnn} \, . \tag{5.26}$$

Note that not all the components of these tensors are independent. For example, the part of $\mathcal{S}$ isotropic in the last two indices is proportional to $\mathcal{A}$, etc. A physical interpretation of the expression (5.23) will be given in section 7 in connection with the continuous-phase momentum equation.

Contrary to (5.7), all these integrals are effected over the surface of a fixed sphere centered at $\mathbf{x}$. This circumstance introduces a significant simplification, some aspects of which will be seen below. It should be noted, however, that an important situation where (5.23) is invalid is the vicinity of boundaries. Here the integral in (5.7) cannot be approximated.

The expressions (5.21), (5.24), (5.25), (5.26) are probably most useful for analytical work. On the other hand, as will be clearer from the considerations of section 10, for numerical computations it appears more convenient to use the same idea underlying the expansion (5.23) in a slightly different way. Now we start from (5.5) rather than (5.7) and write

$$\beta_C (\nabla < f_C > - < \nabla f_C >) = < f_C > \nabla \beta_D + \beta_D \, \mathcal{A}'[f_C] - \nabla \cdot (\beta_D \mathcal{L}'[f_C]) \, , \tag{5.27}$$

where

$$\beta_D \mathcal{A}'[f_C] = \int d^3w \, P(\mathbf{x}, \mathbf{w}; t) \int_{|\mathbf{r}|=a} dS_r \, < f_C >_1 (\mathbf{x} + \mathbf{r}, t|\mathbf{x}, \mathbf{w}) \, \mathbf{n}, \tag{5.28}$$

$$\beta_D \mathcal{L}'[f_C] = \beta_D \mathcal{T}'[f_C] + \nabla \cdot (\beta_D \mathcal{S}'[f_C]) + \nabla\nabla : (\beta_D \mathcal{R}'[f_C]) + \dots . \tag{5.29}$$

As in (5.28), all the primed quantities differ from the corresponding unprimed ones in that the unconditional average $< f_C >$ is not subtracted in the integrand. For certain purposes another convenient form of (5.27) is:

$$\beta_C (\nabla < f_C > - < \nabla f_C >) = \beta_D \, (\mathcal{A}'[f_C] - \nabla < f_C >) - \nabla \cdot [\beta_D \, (\mathcal{L}'[f_C] - \mathbf{I} < f_C >)] \, , \tag{5.30}$$

where $\mathbf{I}$ is the identity two-tensor. The equivalence between these forms and the ones previously given is readily established by expanding $< f_C > (\mathbf{z}, t)$ in a Taylor series centered at $\mathbf{x}$ and carrying out the surface integrations explicitly.

**5.2. Disperse phase.** Starting from the definition (5.1) and by proceeding similarly to the derivation of the expression (3.5) for $\beta_D$, it is easy to show that, for the disperse-phase averages,

$$(5.31)\quad \beta_D < f_D > (\mathbf{x},t) = \int_{|\mathbf{y}-\mathbf{x}|\leq a} d^3y \int d^3w\, P(\mathbf{y},\mathbf{w};t) < f_D >_1 (\mathbf{x},t|\mathbf{y},\mathbf{w}).$$

The integral is over the centers of all the particles that contain the point $\mathbf{x}$.

It is interesting to relate this expression to the particle average previously defined in (4.2). To this end we proceed similarly to (5.19), again exploiting the slow dependence of $< f_D >_1$ with respect to the position $\mathbf{y}$ of the particle center. Setting as before $\mathbf{r} = \mathbf{x} - \mathbf{y}$ and temporarily omitting the explicit indication of the variables irrelevant for this derivation, we write

$$(5.32)\quad \begin{aligned} P(\mathbf{y}) < f_D >_1 (\mathbf{r},\mathbf{y}) = P(\mathbf{x}) < f_D >_1 (\mathbf{r},\mathbf{x}) - \mathbf{r}\cdot \Big\{ &\nabla_x \left[P(\mathbf{x}) < f_D >_1 (\mathbf{r},\mathbf{x})\right] \\ &+\frac{1}{2}\mathbf{r}\cdot\nabla_x\nabla_x \left[P(\mathbf{x}) < f_D >_1 (\mathbf{r},\mathbf{x})\right] \\ &-\frac{1}{6}\mathbf{rr} : \nabla_x\nabla_x\nabla_x \left[P(\mathbf{x}) < f_D >_1 (\mathbf{x},\mathbf{r})\right]\Big\} + O\left(\frac{a^4}{L^4} P < f_D >_1\right). \end{aligned}$$

The number of terms to be retained in this expansion depends on the accuracy desired in $a/L$ and on the order of magnitude of $< f_D >_1$. The terms shown are sufficient for an error of order $(a/L)^2$ when $< f_D >_1 \sim O(L/a)^2$; fewer terms are needed with $< f_D >_1$ is smaller. Upon substitution into (5.31) we then have

$$(5.33)\quad \begin{aligned} \beta_D < f_D > = nv\overline{< f_D >_1} - \nabla\cdot \Big[ &nv\,\overline{\mathbf{r} < f_D >_1} - \frac{1}{2}\nabla\cdot\left(nv\,\overline{\mathbf{rr} < f_D >_1}\right) \\ &+ \frac{1}{6}\nabla\nabla : \left(nv\,\overline{\mathbf{rrr} < f_D >_1}\right)\Big] + O\left(\frac{a^4}{L^4} nv < f_D >_1\right), \end{aligned}$$

where

$$(5.34)\quad \begin{aligned} &\overline{< f_D >_1}(\mathbf{x},t) \\ &= \frac{1}{n}\int d^3w\, P(\mathbf{x},\mathbf{w};t)\left(\frac{1}{v}\int_{|\mathbf{r}|\leq a} d^3r < f_D >_1 (\mathbf{x}+\mathbf{r},t|\mathbf{x},\mathbf{w})\right), \end{aligned}$$

and all the arguments have been restored. Note that, since the inner integral in (5.34) is a quantity referring to the particle as a whole, this expression is analogous to the particle average (4.3). The other overlined quantities in (5.33) are defined similarly. (Admittedly, the notation in the left-hand side of (5.34) is somewhat unsatisfactory as it does not indicate the average of $< f_D >_1$ over the particle volume, but we nevertheless use it for simplicity.)

Upon setting $f_D = 1$, (5.33) reduces to the approximate relation (3.6) between $\beta_D$ and $nv$ previously noted. Another interesting special case of (5.33) is found by taking for $f_D$ the local instantaneous velocity of the disperse phase material $\mathbf{u}_D$. It is evident from the definition of center-of-mass velocity and (5.34) that, for homogeneous particles,

$$\overline{< \mathbf{u}_D >_1} = \frac{1}{n}\int d^3w\, P(\mathbf{x}, \mathbf{w}; t)\, \mathbf{w} = \overline{\mathbf{w}}(\mathbf{x}, t)\,. \tag{5.35}$$

For solid spherical particles

$$\mathbf{u}_D = \mathbf{w} + \boldsymbol{\Omega} \times \mathbf{r}\,, \tag{5.36}$$

where $\boldsymbol{\Omega}$ is the angular velocity around the center. (In some cases, this is an independent variable that must be included in the list of parameters on which $P(N;t)$ depends; this generalization is immediate and does not invalidate the results that follow.) With (5.36), a straightforward calculation gives

$$\nabla \cdot \overline{\mathbf{r} < \mathbf{u}_D >_1} = \frac{1}{5} a^2 \nabla \times (nv\overline{\boldsymbol{\Omega}})\,, \tag{5.37}$$

$$\frac{1}{2}\nabla\nabla : \overline{\mathbf{rr} < \mathbf{u}_D >_1} = \frac{1}{10} a^2 \nabla^2 (nv\overline{\mathbf{w}})\,, \tag{5.38}$$

so that

$$\begin{aligned} &\beta_D < \mathbf{u}_D > \\ &= nv\,\overline{\mathbf{w}} - \frac{1}{5}\, a^2 \nabla \times (nv\,\overline{\boldsymbol{\Omega}}) + \frac{1}{10}\, a^2 \nabla^2 (nv\overline{\mathbf{w}}) + O\left(\frac{a^4}{L^3} nv\,\overline{\boldsymbol{\Omega}}\right). \end{aligned} \tag{5.39}$$

If $\overline{\boldsymbol{\Omega}}$ is driven by the external flow, it will be of the order of the local vorticity in the continuous phase, i.e. of the order of $< \mathbf{u}_C > /L$. In this case then the difference between $< \mathbf{u}_D >$ and $\overline{\mathbf{w}}$ is $O(a^2/L^2)$. This difference could increase to $O(a/L)$ if the particle rotation is driven by external couples or maintained by inertia, for then $\overline{\boldsymbol{\Omega}}$ might be bigger than $O(1/L)$. For non-solid particles (e.g., spherical droplets) $< \mathbf{u}_D >$ and $\overline{\mathbf{w}}$ differ by a term of order $(a/L)$ whenever inertial effects inside the drop are important. If one is willing to live with an error of order $a^2/L^2$ (that can, at times, be as large as $O(a/L)$), $< \mathbf{u}_D >$ and $\overline{\mathbf{w}}$ can be used interchangeably. It should be recognized, however, that for particles with very large aspect ratios, such as fibers, the difference between the two fields can be considerable and more complex relations may be necessary to relate them with acceptable accuracy.

In physical terms, the difference between $< \mathbf{u}_D >$ and $\overline{\mathbf{w}}$ may be clarified by noting that, if $\Delta V$ is a (macroscopic) volume element of the suspension, then $< \mathbf{u}_D > \Delta V$ is the average volume flow rate of the disperse phase due to the motion of all the *particle material* contained inside $\Delta V$. On the other hand, $\overline{\mathbf{w}}\,\Delta V$ is the volume flow rate of the disperse phase due to the motion of all the *particles whose center* is inside $\Delta V$.

The use of particle- in place of phase-averages for the disperse phase has many advantages. For example, as will be shown in section 8, the particle momentum equation can be obtained directly by averaging Newton's law, $m\dot{\mathbf{w}} = \mathbf{f}$, where $m$ is the particle mass and $\mathbf{f}$ the total force. This procedure avoids the need to consider the stress field inside the particles as in the usual two-fluid model derivations (see e.g. Drew 1983), a useful feature for situations in which this stress field is degenerate as, for example, in the case of rigid particles or massless gas bubbles.

Furthermore, particle averages may be useful also e.g. in the case of droplets in which the specification of the internal velocity field requires in principle an infinity of degrees of freedom. For example, for spherical drops, one can expand the velocity field in a suitable orthogonal series, and average the coefficients of the first term of the expansion, of the second term, etc. over all the drops. This would correspond in a sense to a "spectral," rather than pointwise, description of the velocity field and in principle can be carried out exactly. On the other hand, for a situation such as multiphase flow in which approximations are unavoidable, such a description lends itself to more fruitful approximations than the pointwise one. For example, one might want to retain only the first few terms of the expansion thus capturing the salient features of the internal velocity distribution. This remark illustrates the ability of the method to limit the number of degrees of freedom used in the description of the disperse phase tailoring it to the specific situation at hand.

In some cases, one may also expect that the use of $\overline{\mathbf{w}}$ in place of $< \mathbf{u}_D >$ might lead to simpler equations. For example, for rigid spheres in inviscid flow, particle rotation cannot have any dynamical significance and it is then evident from (5.39) that an equation phrased in terms of $< \mathbf{u}_D >$ must contain terms that cancel the effect of rotation and therefore be more complex than one in terms of $\overline{\mathbf{w}}$.

**6. Conditional averages.** It has been shown in the previous section that the evaluation of derivatives of the average quantities requires a knowledge of the conditionally averaged fields. Herein lies the root of the well-known closure problem, as equations for these averages involve averages of still higher order and so forth. Nevertheless it is important to derive averaged equations for the conditionally averaged fields. In the first place, they provide an insight into the structure of the theory. Secondly, they are necessary to close the equations in the dilute limit and as a starting point for several approximations.

We consider continuous phase averages first. Starting from the definition (5.3) of conditional average, for the spatial derivatives of $< f_C >_K$ we have

$$\nabla \left(\beta_C^{(K)} < f_C >_K\right) = \beta_C^{(K)} < \nabla f_C >_K + \frac{1}{(N-K)!} \int d\mathcal{C}^{N-K}\, P(N-K|K;t)\, f_C\, \nabla \chi_C . \tag{6.1}$$

Proceeding as done before to obtain (5.5), and omitting explicit indication of the dependence on time and velocity variables for simplicity of writing, we find

$$\begin{aligned}\nabla(\beta_C^{(K)} < f_C >_K)(\mathbf{x}|K) = \; & \beta_C^{(K)} < \nabla f_C >_K + \sum_{\alpha=1}^{K} \delta\left(a - |\mathbf{x} - \mathbf{y}^\alpha|\right) \mathbf{n}^\alpha < f_C >_K (\mathbf{x}|K) \\ & + \int_{|\mathbf{x}-\mathbf{y}^{K+1}|=a} dS^{K+1} \mathbf{n}^{K+1} \int d^3w\, P\left(\mathbf{y}^{K+1}|K\right) < f_C >_{K+1} (\mathbf{x}|K+1),\end{aligned} \tag{6.2}$$

where $P\left(\mathbf{y}^{K+1}|K\right) = P(K+1)/P(K)$. With $f_C = 1$ we find from this expression

$$\begin{aligned}\nabla \beta_C^{(K)}(\mathbf{x}|K) = \; & \sum_{\alpha=1}^{K} \delta\left(a - |\mathbf{x} - \mathbf{y}^\alpha|\right) \mathbf{n}^\alpha \\ & + \int_{|\mathbf{x}-\mathbf{y}^{K+1}|=a} dS^{K+1} \mathbf{n}^{K+1} \int d^3w\, P\left(\mathbf{y}^{K+1}|K\right) .\end{aligned} \tag{6.3}$$

This relation can be used in (6.2) to put the result in a form similar to (5.7), namely

$$\begin{aligned}\nabla < f_C >_K = < \nabla f_C >_K & + \frac{1}{\beta_C^{(K)}} \int_{|\mathbf{x}-\mathbf{y}^{K+1}|=a} dS^{K+1} \mathbf{n}^{K+1} \\ & \times \int d^3w\, P\left(\mathbf{y}^{K+1}|K\right) \left[< f_C >_{K+1} (\mathbf{x}|K+1) - < f_C >_K (\mathbf{x}|K)\right] .\end{aligned} \tag{6.4}$$

The calculation of the time derivative is not as straightforward because this quantity depends on time not only directly, for fixed $\mathbf{x}$ and $\mathcal{C}^K$, but also through the evolution of the configuration of the first $K$ particles. In the applications that follow one needs the time derivative of $< f_C >_K$ at a certain time, given $\mathcal{C}^K$ at that time. Where the first $K$ particles are at $t + dt$ is irrelevant for this purpose and therefore one must allow for all the possibilities. This argument suggests the following definition:

$$\begin{aligned}\frac{\partial}{\partial t}(\beta_C^{(K)} < f_C >_K) \equiv \; & \frac{1}{(N-K)!} \int d\mathcal{C}^{N-K} \left[ \frac{\hat{\partial}}{\hat{\partial} t} + \sum_{\alpha=1}^{K} (\mathbf{w}^\alpha \cdot \nabla_\alpha \right. \\ & \left. + \dot{\mathbf{w}}^\alpha \cdot \boldsymbol{\Delta}_\alpha) \right] \left[\chi_C(\mathbf{x}; N)\, f_C(\mathbf{x}, t; N)\, P(N-K|K;t)\right] .\end{aligned} \tag{6.5}$$

Executing the indicated operations, it is not difficult to find

$$\frac{\partial}{\partial t}(\beta_C^{(K)} < f_C >_K) =$$

$$\beta_C^{(K)} \left\langle \frac{\partial f_C}{\partial t} \right\rangle_K + \frac{\beta_C^{(K)}}{P(K;t)} \sum_{\alpha=1}^{K} \langle f_C \, \Delta_\alpha \cdot [(\ll \dot{\mathbf{w}}^\alpha \gg_K - \dot{\mathbf{w}}^\alpha) \; P(K;t)] \rangle_K$$

$$+ \frac{1}{(N-K)!} \sum_{\alpha=1}^{N} \int dC^{N-K} \, f_C \, P(N-K|K;t) \, \mathbf{w}^\alpha \cdot \nabla_\alpha \chi_C \,, \tag{6.6}$$

where the partial derivative in the first term in the right-hand side is as defined in (5.9).

Upon using (6.1) and (6.6), and recalling the relation (3.4) for $\nabla \chi_C$ and the kinematic boundary condition (5.14), we can write down the analogue of the transport theorem (5.15):

$$\frac{\partial}{\partial t}(\beta_C^{(K)} < f_C >_K) + \nabla \cdot (\beta_C^{(K)} < f_C \mathbf{u}_C >_K) = \beta_C^{(K)} \left\langle \frac{\partial f_C}{\partial t} + \nabla \cdot (f_C \mathbf{u}_C) \right\rangle_K$$

$$+ \frac{\beta_C^{(K)}}{P(K;t)} \sum_{\alpha=1}^{K} \langle f_C \, \Delta_\alpha \cdot [(\ll \dot{\mathbf{w}}^\alpha \gg_K - \dot{\mathbf{w}}^\alpha) \; P(K;t)] \rangle_K \,. \tag{6.7}$$

This relation shows that the mean transport (left-hand side) is due to the microscopic transport (first term in the right-hand side) plus a contribution due to the fact that the particle velocities differ from their mean value. For example, upon taking $f_C = 1$, we have

$$\frac{\partial \beta_C^{(K)}}{\partial t} + \nabla \cdot (\beta_C^{(K)} < \mathbf{u}_C >_K) \;=\; \beta_C^{(K)} \langle \nabla \cdot \mathbf{u}_C \rangle_K$$

$$+ \frac{\beta_C^{(K)}}{P(K;t)} \sum_{\alpha=1}^{K} \langle \Delta_\alpha \cdot [(\ll \dot{\mathbf{w}}^\alpha \gg_K - \dot{\mathbf{w}}^\alpha) \; P(K;t)] \rangle_K \,. \tag{6.8}$$

Similar relations can be worked out for the disperse-phase conditional averages. Since they will not be used in the following, we do not present them here.

For conditional particle averages, transport theorems similar to (4.6) are of interest. To avoid unnecessary complications, we present a derivation for the simple example introduced in section 4.

We encounter here the same ambiguity mentioned before in connection with the time derivative of $< f_C >_K$. For the same reasons given in that connection, on the basis of (4.8), we define this derivative by

$$\frac{\partial}{\partial t}\left(n^{(1)} \overline{g}^{(1)}\right)(\mathbf{x}, t|\mathbf{z}) \;=\; \frac{1}{(N-2)!}$$

$$\times \int dC^{N-2} \left( \frac{\hat{\partial}}{\hat{\partial} t} + \mathbf{w}^z \cdot \nabla_z \right) \left[ P(\mathbf{x}, N-2; t|\mathbf{z}) \, g^1(\mathbf{x}, \mathbf{z}, N-2; t) \right], \tag{6.9}$$

where $\mathbf{w}^z$ is the velocity of the particle at $\mathbf{z}$. Upon using the evolution equation (2.17) for the conditional probability and the divergence theorem to integrate terms with $\alpha = 3, 4, \ldots, N$, one readily finds

$$\begin{aligned} &\frac{\partial}{\partial t}\left(n^{(1)}\overline{g}^{(1)}\right) + \nabla \cdot \left[n^{(1)}\overline{\mathbf{w}g}^{(1)}\right] = \\ &n^{(1)}\overline{\dot{g}^1}^{(1)} + \frac{n^{(1)}}{P(\mathbf{z};t)}\overline{g^1\nabla_z \cdot \left[\left(\ll \mathbf{w}^z \gg_1 - \mathbf{w}^z\right) P(\mathbf{z};t)\right]}^{(1)}, \end{aligned} \tag{6.10}$$

where $\dot{g}^1$ is as defined in (4.4).

By following the same principles, these results can readily be extended to situations in which the velocities are independent variables and other conditional averages are of interest.

**7. Continuous-phase momentum.** We assume the continuous phase to be incompressible and for its description we use the phase averages (5.1). Thus, from (5.17), we find the continuity equation in the form

$$\frac{\partial \beta_C}{\partial t} + \nabla \cdot (\beta_C < \mathbf{u}_C >) = 0. \tag{7.1}$$

The exact, unaveraged momentum equation is

$$\frac{\partial\, \rho_C \mathbf{u}_C}{\partial t} + \nabla \cdot (\rho_C \mathbf{u}_C \mathbf{u}_C) = \nabla \cdot \boldsymbol{\sigma}_C + \rho_C \mathbf{g}, \tag{7.2}$$

where $\rho_C$ is the density, $\mathbf{g}$ is the body force, and $\boldsymbol{\sigma}_C$ is the stress tensor. Upon using the transport theorem (5.15), the averaged form is

$$\frac{\partial}{\partial t}(\beta_C \rho_C < \mathbf{u}_C >) + \nabla \cdot (\rho_C \beta_C < \mathbf{u}_C \mathbf{u}_C >) = \beta_C < \nabla \cdot \boldsymbol{\sigma}_C > + \rho_C \beta_C \mathbf{g}. \tag{7.3}$$

The spatial differentiation relation (5.20) gives

$$\beta_C < \nabla \cdot \boldsymbol{\sigma}_C > = \beta_C \nabla \cdot < \boldsymbol{\sigma}_C > - \beta_D \mathcal{A}[\boldsymbol{\sigma}_C] + \nabla \cdot (\beta_D \mathcal{L}[\boldsymbol{\sigma}_C]), \tag{7.4}$$

so that the momentum equation (7.3) becomes

$$\begin{aligned} &\frac{\partial}{\partial t}(\beta_C \rho_C < \mathbf{u}_C >) + \nabla \cdot (\rho_C \beta_C < \mathbf{u}_C \mathbf{u}_C >) = \\ &\beta_C \nabla \cdot < \boldsymbol{\sigma}_C > - \beta_D \mathcal{A}[\boldsymbol{\sigma}_C] + \nabla \cdot (\beta_D \mathcal{L}[\boldsymbol{\sigma}_C]) + \rho_C \beta_C \mathbf{g}. \end{aligned} \tag{7.5}$$

Recalling the definitions (5.21) and (5.22) of $\mathcal{A}$ and $\mathcal{L}$, we see that the process of interchanging the order of averaging and differentiation in the stress term results in the explicit appearance of the conditionally averaged stress field in these quantities.

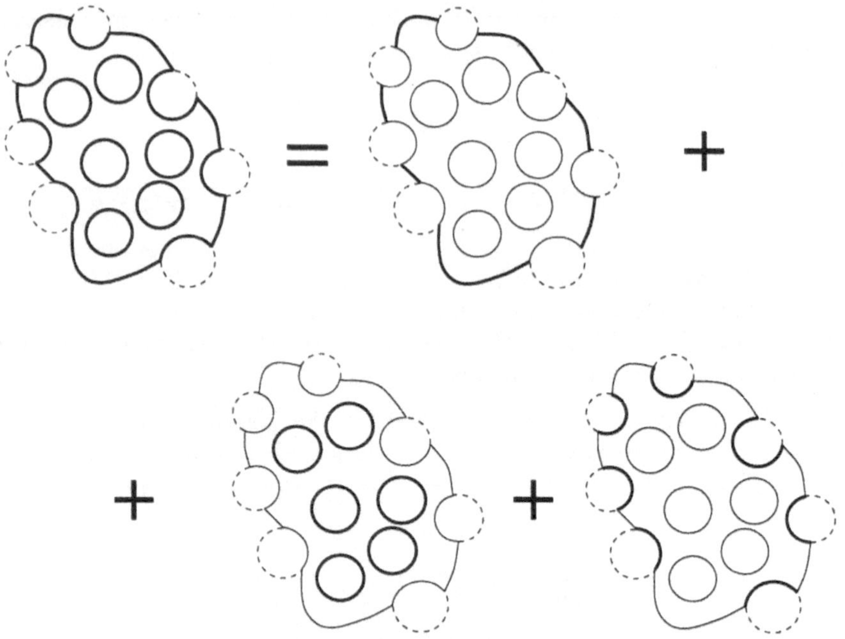

FIG. 2. *A schematic representation of the spatial differentiation rule (5.20) or (7.4). See text after Eq. (7.5).*

The physical meaning of the relation (7.4) can be illustrated with reference to Fig. 2, which refers to a (macroscopic) volume element $\Delta V$ of the mixture. Then $(\beta_C < \nabla \cdot \boldsymbol{\sigma}_C >) \Delta V = < \nabla \cdot \boldsymbol{\sigma}_C > \Delta V_C$ is the total stress exerted on the continuous phase at the interfaces shown by the heavy lines. These include the surface of $\Delta V$ in the continuous phase, the surfaces of the particles entirely contained in $\Delta V$, and those of the particles crossed by the boundary of $\Delta V$ and only partially contained in $\Delta V$. The first term in the right-hand side, $(\beta_C \nabla \cdot < \boldsymbol{\sigma}_C >) \Delta V$ is the stress transmitted through the surface of $\Delta V$ in contact with the continuous phase. The second term, $(\beta_D \mathcal{A}[\boldsymbol{\sigma}_C]) \Delta V$ has the nature of a momentum source, and corresponds to the stress exerted on the continuous phase by the particles entirely within $\Delta V$. The final term, $\nabla \cdot (\beta_D \mathcal{L}[\boldsymbol{\sigma}_C]) \Delta V$ accounts for the stress exerted at the surface of the particles partially included in $\Delta V$. The three-term truncation (5.23) corresponds to approximating $\mathcal{L}[\boldsymbol{\sigma}_C]$ as a surface stress acting at the portion of the boundary of $\Delta V$ cutting through the particles.

Upon introducing the "kinematic" Reynolds stress $\mathbf{M}_C$ by the definition

$$\begin{aligned}\mathbf{M}_C &= <\mathbf{u}_C><\mathbf{u}_C> - <\mathbf{u}_C\mathbf{u}_C> \\ &= -<(\mathbf{u}_C - <\mathbf{u}_C>)(\mathbf{u}_C - <\mathbf{u}_C>)>,\end{aligned} \tag{7.6}$$

and using the continuity equation (7.1), we can rewrite the momentum equation (7.5) in the form

$$\begin{aligned}&\rho_C \beta_C \left(\frac{\partial <\mathbf{u}_C>}{\partial t} + <\mathbf{u}_C> \cdot \nabla <\mathbf{u}_C>\right) = \\ &\beta_C \nabla \cdot <\boldsymbol{\sigma}_C> - \beta_D \mathcal{A}[\boldsymbol{\sigma}_C] + \nabla \cdot (\beta_C \rho_C \mathbf{M}_C + \beta_D \mathcal{L}[\boldsymbol{\sigma}_C]) + \rho_C \beta_C \mathbf{g}.\end{aligned} \tag{7.7}$$

The terms in the first line have the structure that is expected intuitively: the volume occupied by the continuous-phase per unit mixture volume is $\beta_C$ and the particles act as a source of momentum of the order of their number density times the "extra" force per particle $\mathcal{A}$. The Reynolds transport term is similarly expected. The term $\mathcal{L}$, showing that the particles act not only as a source of momentum, but also contribute to its transport, is perhaps less obvious. The argument given before (7.6) shows that its physical origin is rooted in the finite size of the particles.

For an incompressible, Newtonian continuous phase, $\boldsymbol{\sigma}_C = -p_C \mathbf{I} + 2\mu_C \mathbf{e}_C$, where $p_C$ is the pressure, $\mu_C$ the viscosity coefficient, and $\mathbf{e}_C = \frac{1}{2}[\nabla \mathbf{u}_C + (\nabla \mathbf{u}_C)^T]$ the strain rate and therefore $<\boldsymbol{\sigma}_C> = -<p_C> \mathbf{I} + 2\mu_C <\mathbf{e}_C>$. To derive an expression for the mean strain rate, let us first define the mixture volumetric flow rate $\mathbf{u}_m$ by

$$\mathbf{u}_m = \beta_C <\mathbf{u}_C> + \beta_D <\mathbf{u}_D>. \tag{7.8}$$

Then, from (5.5), its analogue for the disperse phase, and the no slip condition (or, more generally, continuity of the velocity at the particle interface), one finds

$$\nabla \mathbf{u}_m = \beta_C <\nabla \mathbf{u}_C> + \beta_D <\nabla \mathbf{u}_D>. \tag{7.9}$$

On the basis of this relation, the mean rate of strain of the suspension $\mathbf{E}_m$, defined by

$$\mathbf{E}_m = \frac{1}{2}\left[\nabla \mathbf{u}_m + (\nabla \mathbf{u}_m)^T\right], \tag{7.10}$$

is therefore given by

$$\mathbf{E}_m = \beta_C <\mathbf{e}_C> + \beta_D <\mathbf{e}_D>. \tag{7.11}$$

For rigid particles the strain rate $\mathbf{e}_D$ vanishes identically so that

$$<\mathbf{e}_C> = \frac{1}{\beta_C}\mathbf{E}_m. \tag{7.12}$$

For fluid particles the situation is more complicated and one has to use the general differentiation formulae given before in section 5. In this way one finds, correct to order $(a/L)$,

$$\begin{aligned}(7.13)\qquad <\mathbf{e}_C>(\mathbf{x},t) = \mathbf{E}_C - \frac{1}{\beta_C}\mathrm{Symm}\Big\{\beta_D\mathcal{A}[\mathbf{u}_C] \\ - \nabla\cdot(\beta_D\mathcal{T}[\mathbf{u}_C]) - \nabla\nabla : (\beta_D\mathcal{S}[\mathbf{u}_C])\Big\},\end{aligned}$$

where Symm denotes the symmetric part of a second-order tensor and $\mathbf{E}_C$ is the strain rate of the average velocity field:

$$(7.14)\qquad \mathbf{E}_C = \mathrm{Symm}\{\nabla < \mathbf{u}_C >\}.$$

It can readily be shown by using the velocity representation (5.36) that, for rigid particles, (7.13) reduces to (7.12).

Since $\mathcal{A}[\boldsymbol{\sigma}_C]$ and $\mathcal{L}[\boldsymbol{\sigma}_C]$ contain the conditionally averaged stresses $< \boldsymbol{\sigma}_C >_1$, to close the problem, one may attempt to consider the conditionally averaged equations. By proceeding as before with the help of (6.7) one finds

$$\begin{aligned}(7.15)\qquad \frac{\partial}{\partial t}(\beta_C^{(1)} < \mathbf{u}_C >_1) + \nabla\cdot(\beta_C^{(1)} < \mathbf{u}_C\mathbf{u}_C >_1) = \frac{\beta_C^{(1)}}{\rho_C} < \boldsymbol{\nabla}\cdot\boldsymbol{\sigma}_C >_1 \\ + \beta_C^{(1)}\mathbf{g} + \frac{\beta_C^{(1)}}{P(1;t)}\langle \mathbf{u}_C\,\Delta\cdot[(\dot{\mathbf{w}} - \ll \dot{\mathbf{w}} \gg_1)\,P(1;t)]\,\rangle_1,\end{aligned}$$

This equation differs in form from the unconditional average equation (7.3) given before due to the presence of the last term. Note that, in the dilute limit $\beta_D \to 0$, the difference $\dot{\mathbf{w}} - \ll \dot{\mathbf{w}} \gg_1$ tends to zero as the system becomes then totally deterministic.

We have written (7.15) similarly to (7.3) in that $< \boldsymbol{\nabla}\cdot\boldsymbol{\sigma}_C >_1$ rather than $\boldsymbol{\nabla}\cdot < \boldsymbol{\sigma}_C >_1$ appears in the equation. As before, the process of interchanging averaging and differentiation causes the explicit appearance of the two-particle averaged stress $< \boldsymbol{\sigma}_C >_2$. This step however is not as direct as in the case of the unconditionally averaged fields. Since we only have a limited use for (7.15), we do not show the form explicitly.

**8. Disperse-phase momentum.** A result analogous to (5.17) also holds for the disperse phase. Again assuming incompressibility we thus find, as in (7.1),

$$(8.1)\qquad \frac{\partial \beta_D}{\partial t} + \nabla\cdot(\beta_D < \mathbf{u}_D >) = 0.$$

As argued before, however, it is advantageous to describe the continuous phase in terms of particle, rather than phase, averages, and therefore we prefer to use the conservation equation (4.6) for the particle number density in place of (8.1). Of course, if the particle volume remains constant and the approximations $\beta_D \simeq nv$, $< \mathbf{u}_D > \simeq \overline{\mathbf{w}}$ are acceptable, the two equations reduce to each other.

One of the significant advantages of using the particle average is evident in the derivation of the disperse-phase equation of motion. The exact equation of motion for the particle with its center at $\mathbf{x}$ is

$$m\,\dot{\mathbf{w}}(\mathbf{x},t) = \int_{|\mathbf{r}|=a} dS_r\, \boldsymbol{\sigma}_C(\mathbf{x}+\mathbf{r},t;N)\cdot\mathbf{n} + \mathbf{f}_c + m\,\mathbf{g}\,, \tag{8.2}$$

where $\mathbf{f}_c$ is the collision force. Upon averaging this equation according to (4.2) and use of the transport theorem (5.15), one finds for particles with equal mass

$$\begin{aligned} & m\,\frac{\partial\, n\,\overline{\mathbf{w}}}{\partial t} + m\nabla\cdot(n\overline{\mathbf{w}\mathbf{w}}) = \\ & \int d^3w P(\mathbf{x},\mathbf{w},t) \int_{|\mathbf{x}-\mathbf{z}|=a} dS_z\, <\boldsymbol{\sigma}_C>_1 \cdot\mathbf{n} + \nabla\cdot\boldsymbol{\sigma}_c + n\,m\,\mathbf{g}\,. \end{aligned} \tag{8.3}$$

The term $\boldsymbol{\sigma}_c$ is the collision stress due to direct particle-particle interaction derived in Appendix C of Zhang and Prosperetti (1997; see also Sangani and Didwania 1993a; Zhang and Rauenzahn 1997). It should be noted that the derivation of this term does not presuppose binary or short-duration collisions.

To put Eq. (8.3) in a more useful form, we note that

$$\begin{aligned} & \int d^3w\, P(\mathbf{x},\mathbf{w},t) \int_{|\mathbf{x}-\mathbf{z}|=a} dS_z\, <\boldsymbol{\sigma}_C>_1 \cdot\mathbf{n} \\ & = \beta_D \mathcal{A}[\boldsymbol{\sigma}_C] + \int d^3w P(\mathbf{x},\mathbf{w},t)\int_{|\mathbf{r}|=a} dS_r\, <\boldsymbol{\sigma}_C>(\mathbf{x}+\mathbf{r},t)\cdot\mathbf{n} \\ & = \beta_D \mathcal{A}[\boldsymbol{\sigma}_C] + nv\nabla\cdot <\boldsymbol{\sigma}_C>(\mathbf{x},t) + O\left(\beta_D\left(\frac{a}{L}\right)^2 <\boldsymbol{\sigma}_C>\right), \end{aligned} \tag{8.4}$$

where $\mathcal{A}[\boldsymbol{\sigma}_C]$ is defined in (5.21) and $< \boldsymbol{\sigma}_C > (\mathbf{z},t)$ has been expanded in a Taylor series around $\mathbf{x}$ thereby introducing an error of order $(a^2/L^2)$. Dropping an error of the same order arising from the approximation $\beta_D \simeq nv$, we may then rewrite the momentum equation (8.3) as

$$\begin{aligned} & \rho_D\beta_D\left(\frac{\partial\,\overline{\mathbf{w}}}{\partial t} + \overline{\mathbf{w}}\cdot\nabla\overline{\mathbf{w}}\right) = \\ & \beta_D\nabla\cdot < \boldsymbol{\sigma}_C > + \beta_D\mathcal{A}[\boldsymbol{\sigma}_C] + \rho_D\nabla\cdot(\beta_D\mathbf{M}_D) + \nabla\cdot\boldsymbol{\sigma}_c + \rho_D\beta_D\mathbf{g}\,, \end{aligned} \tag{8.5}$$

where $\rho_D$ is the density of the particle material. Here we have used the number density conservation equation (4.6) and introduced the kinematic Reynolds stress for the disperse phase $\mathbf{M}_D$ defined by

$$\mathbf{M}_D = \overline{\mathbf{w}}\,\overline{\mathbf{w}} - \overline{\mathbf{w}\mathbf{w}} = -\overline{(\mathbf{w}-\overline{\mathbf{w}})(\mathbf{w}-\overline{\mathbf{w}})}\,. \tag{8.6}$$

It will be noted that (8.5) explicitly shows the particles to respond to the continuous-phase stress. This feature is according to physical intuition. The disperse-phase stress – e.g. the pressure inside the particles – cannot affect the motion of the particles directly, but only indirectly through its relation with the continuous-phase stress resulting from dynamical boundary conditions at the particles surface (see Prosperetti and Zhang 1997). This rather involved conceptual circle is avoided by the above procedure. Furthermore, there are cases where a disperse-phase pressure cannot even be meaningfully defined, as for rigid particles (see e.g. Givler 1987). Of course, Eq. (8.5) does not imply that further information on the internal dynamics of the particles is not necessary, as this will be needed to calculate the conditionally-averaged fields. However, the type of information required can be tailored to the problem. For example, the case of rigid particles can be addressed directly rather than as the limit of stiffer and stiffer particles (Drew and Lahey 1993). Similarly, in the case of gas bubbles, there is no need to solve the momentum equation in the gas, but only to state, for example, that the gas pressure is spatially uniform inside each bubble.

The relation (8.5) expresses the momentum balance for all the particles whose center is inside a control volume. An alternative momentum equation giving the momentum balance for all the particle material entirely inside the control volume can be obtained by performing the phase average according to (5.1). While this calculation could be constructed so as to closely parallel the analysis of the continuous-phase momentum equation and lead to a form in which the average acceleration of the disperse phase is determined by the average stress of the disperse phase, this method has the disadvantage of requiring the consideration of the particle constitutive relation. For this reason we follow a slightly different path in which the dynamic boundary condition at the particle surface is invoked early on to eliminate the stress tensor of the particle material. Body forces are neglected for simplicity.

We start from the unaveraged momentum equation for the particle material, assumed incompressible as before:

$$\rho_D \mathbf{a}_D \equiv \frac{\partial \rho_D \mathbf{u}_D}{\partial t} + \nabla \cdot (\rho_D \mathbf{u}_D \mathbf{u}_D) = \nabla \cdot \boldsymbol{\sigma}_D \,, \tag{8.7}$$

where $\mathbf{a}_D$ and $\boldsymbol{\sigma}_D$ are the acceleration and stress of the disperse phase material. After application of the disperse-phase transport theorem (5.16), one finds

$$\frac{\partial}{\partial t}(\beta_D \rho_D < \mathbf{u}_D >) + \nabla \cdot (\beta_D \rho_D < \mathbf{u}_D \mathbf{u}_D >) = \beta_D < \nabla \cdot \boldsymbol{\sigma}_D > + \beta_D \rho_D \mathbf{g}. \tag{8.8}$$

From (5.33) we may write

$$\begin{aligned}\beta_D <\nabla\cdot\boldsymbol{\sigma}_D> = \int d^3w P(\mathbf{x},\mathbf{w};t)\int_{|\mathbf{z}-\mathbf{x}|\leq a} d^3z <\nabla_z\cdot\boldsymbol{\sigma}_D>_1(\mathbf{z},t|\mathbf{x},\mathbf{w})\\ +\nabla\cdot\boldsymbol{\Sigma}_a + O\left(\frac{a^2}{L^2}\beta_D <\nabla\cdot\boldsymbol{\sigma}_D>_1\right),\end{aligned} \tag{8.9}$$

where the second-order tensor $\boldsymbol{\Sigma}_a$ is given by

$$\begin{aligned}\boldsymbol{\Sigma}_a &= -\int d^3w\, P(\mathbf{x},\mathbf{w})\int_{|\mathbf{r}|\leq a} d^3r < \nabla_r\cdot\boldsymbol{\sigma}_D >_1(\mathbf{x}+\mathbf{r},t|\mathbf{x},\mathbf{w})\,\mathbf{r}\\ &= -\rho_D\int d^3wP(\mathbf{x},\mathbf{w})\int_{|\mathbf{r}\leq a} d^3r < \mathbf{a}_D >_1(\mathbf{x}+\mathbf{r},t|\mathbf{x},\mathbf{w})\,\mathbf{r}\,.\end{aligned} \tag{8.10}$$

The first term in the right-hand side of (8.9) can be further manipulated by interchanging the conditional averaging and the integration over the particle volume and then applying the divergence theorem to write it as an integral over the particle surface. Upon using the dynamic boundary condition

$$\boldsymbol{\sigma}_D\cdot\mathbf{n} = \boldsymbol{\sigma}_C\cdot\mathbf{n}\,, \tag{8.11}$$

we then have

$$\begin{aligned}\int d^3wP(\mathbf{x},\mathbf{w};t)\int_{|\mathbf{z}-\mathbf{x}|\leq a} d^3z < \nabla_z\cdot\boldsymbol{\sigma}_D >_1 =\\ \int d^3wP(\mathbf{x},\mathbf{w};t)\int_{|\mathbf{x}-\mathbf{z}|=a} dS_z < \boldsymbol{\sigma}_C >_1\cdot\mathbf{n}\end{aligned} \tag{8.12}$$

Finally, by (8.4), we find the following form for the disperse phase momentum equation:

$$\begin{aligned}\rho_D\left[\frac{\partial\beta_D < \mathbf{u}_D >}{\partial t} + \nabla\cdot(\beta_D < \mathbf{u}_D\mathbf{u}_D >)\right] =\\ \beta_D\nabla\cdot <\boldsymbol{\sigma}_C> +\beta_D\mathcal{A}[\boldsymbol{\sigma}_C] + \nabla\cdot\boldsymbol{\Sigma}_a + \rho_D\beta_D\mathbf{g} + O\left(\tfrac{a^2}{L^2}\beta_D <\nabla\cdot\boldsymbol{\sigma}_D>_1\right).\end{aligned} \tag{8.13}$$

This derivation has been given neglecting surface tension and collision forces. It can be shown, however, that the result is valid in general (Zhang and Prosperetti 1997).

It is interesting to compare (8.13) with the momentum equation (8.5) in terms of $\overline{\mathbf{w}}$. To this end we use once again the transport theorem (5.16) and find

$$\begin{aligned}\rho_D\left[\frac{\partial\beta_D < \mathbf{u}_D >}{\partial t} + \nabla\cdot(\beta_D < \mathbf{u}_D\mathbf{u}_D >)\right] =\\ \int_{|\mathbf{x}-\mathbf{y}|\leq a} d^3y\int d^3wP(\mathbf{y},\mathbf{w};t) < \rho_D\mathbf{a}_D >_1(\mathbf{x},t|\mathbf{y},\mathbf{w})\,.\end{aligned} \tag{8.14}$$

We now effect a translation similar to (5.19) to obtain

$$
\rho_D \left[ \frac{\partial \beta_D < \mathbf{u}_D >}{\partial t} + \nabla \cdot (\beta_D < \mathbf{u}_D \mathbf{u}_D >) \right] =
$$

$$
\int d^3 w \, P(\mathbf{x}, \mathbf{w}; t) \left\langle \int_{|\mathbf{r}| \leq a} d^3 r \rho_D \mathbf{a}_D(\mathbf{x} + \mathbf{r}, t | \mathbf{x}, \mathbf{w}) \right\rangle_1 + \nabla \cdot \mathbf{\Sigma}_a \,. \tag{8.15}
$$

The inner integral in the right-hand side obviously equals the particle mass times its center-of-mass acceleration $\dot{\mathbf{w}}$ so that we have

$$
\rho_D \left[ \frac{\partial \beta_D < \mathbf{u}_D >}{\partial t} + \nabla \cdot (\beta_D < \mathbf{u}_D \mathbf{u}_D >) \right] = nm\overline{\dot{\mathbf{w}}} + \nabla \cdot \mathbf{\Sigma}_a \,. \tag{8.16}
$$

Upon using this relation to express the left-hand side of (8.13), one recovers (8.5).

This argument is consistent with the earlier statement that the description in terms of $\overline{\mathbf{w}}$ is not an approximation but is just as legitimate as the one in terms of $< \mathbf{u}_D >$. The difference between (8.5) and (8.13) is due to the fact that the two refer to slightly different systems. However it may be expected that, in most cases in which a description in terms of averaged quantities is justified, the difference between the two descriptions will be small.

An equation for the average angular momentum of the particles can also be obtained in a manner similar to (8.5). Among other reasons of interest, this relation highlights the physical reality of the interaction expressed by the tensor $\mathcal{T}[\boldsymbol{\sigma}_C]$. If the inertia of the rotational motion is considered, the phase space must be extended to include the angular velocities $\mathbf{\Omega}^\alpha$ of the particles. This introduces some straightforward changes in the previous relations. For example, the integrals $\int d^3 w$ should be changed to $\int d^3 w \int d^3 \Omega$.

The angular momentum equation for each particle is

$$
J \frac{d\mathbf{\Omega}}{dt} = \int_{|\mathbf{x}-\mathbf{z}|=a} dS_z \, (\mathbf{z} - \mathbf{x}) \times (\boldsymbol{\sigma}_C \cdot \mathbf{n}) + \mathbf{L} + \mathbf{m}_c \,, \tag{8.17}
$$

where $J$ is the moment of inertia, $\mathbf{L}$ the external torque, and $\mathbf{m}_c$ the torque due to collisions. Upon averaging this equation according to (4.2) and following the same procedure used to derive (8.3), we find

$$
J \left[ \frac{\partial n \overline{\Omega}}{\partial t} + \nabla (n \overline{\mathbf{\Omega} \mathbf{w}}) \right] =
$$

$$
a \int d^3 w \int d^3 \Omega \, P(\mathbf{x}, \mathbf{w}, \mathbf{\Omega}; t) \int_{|\mathbf{x}-\mathbf{z}|=a} dS_z \, \mathbf{n} \times (< \boldsymbol{\sigma}_C >_1 \cdot \mathbf{n}) + n \overline{\mathbf{L}} + \overline{\mathbf{m}}_c . \tag{8.18}
$$

The integral in the right-hand side can be connected to the skew-symmetric part of $\mathcal{T}[\boldsymbol{\sigma}_C]$ similarly to the calculation done above in (8.4). The result is

$$a\,\epsilon_{ijk}\int d^3w \int d^3\Omega P(\mathbf{x},\mathbf{w},\boldsymbol{\Omega};t)\int_{|\mathbf{x}-\mathbf{z}|=a} dS_z\,(<\boldsymbol{\sigma}_C>_1\cdot\mathbf{n})_j\,n_k \tag{8.19}$$
$$= n\,v\,\epsilon_{ijk}(\mathcal{T}[\boldsymbol{\sigma}_C])_{jk} + O\left(\frac{a}{L}\right)^2 .$$

With the neglect of the error term and use of the continuity equation (4.6), the average angular momentum balance for the disperse phase becomes

$$nJ\left[\frac{\partial\overline{\Omega}_i}{\partial t} + \overline{\mathbf{w}}\cdot\nabla\overline{\Omega}_i\right] = -\beta_D\,\epsilon_{ijk}(\mathcal{T}[\boldsymbol{\sigma}_C])_{jk} + n\overline{L}_i + nJ\frac{\partial M_{\Omega ij}}{\partial x_j} + \overline{m}_{ci}, \tag{8.20}$$

in which

$$\mathbf{M}_\Omega = \overline{\boldsymbol{\Omega}}\,\overline{\mathbf{w}} - \overline{\boldsymbol{\Omega}\mathbf{w}}\,. \tag{8.21}$$

Since torques do not satisfy an action-reaction principle, unlike the linear momentum equation (8.3), the mean torque due to collisions cannot be written as a divergence. It can, however, be expressed as the skew-symmetric part of $\boldsymbol{\sigma}_c$.

**9. Dilute limit.** The momentum equations (7.7) and (8.5) and the continuity equations (7.1) and (4.6) constitute a general two-fluid model for disperse two-phase flows. To close the system one needs constitutive relations expressing the integrals $\mathcal{A}$, $\mathcal{T}$, etc. in terms of average quantities. In principle these relations should be obtained by solving appropriate equations for the one-particle conditionally averaged fields which however, as was remarked at the beginning of section 5, involve other higher-order conditionally averaged fields and are plagued by divergent integrals (see Hinch 1977; Sangani 1991). The only simple case is the dilute limit, with results correct to first order in the particle volume fraction $\beta_D$, that we address in this section. For the viscous case, it is shown in Zhang (1997) that the process can be continued to the next order in $\beta_D$ reducing the problem to that studied by Hinch (1977). It appears doubtful that much progress can be made analytically beyond these results. A more promising approach seems to be the direct calculation of the integrals by numerical simulation. This point will be briefly considered in section 10.

Schematically, the reason why the first order in $\beta_D$ is tractable can be explained in the following terms. Consider for example the continuous-phase momentum equation (7.7) in which we write $<\mathbf{u}_C\mathbf{u}_C> = <\mathbf{u}_C><\mathbf{u}_C> - \mathbf{M}_C$ and note that, for laminar flows, $\mathbf{M}_C = 0$ in the absence of particles. If all terms that vanish for $\beta_D \to 0$ are omitted from (7.7), one then finds that $<\mathbf{u}_C>$, $<p_C>$ satisfy the usual form of the Navier-Stokes equation for a pure fluid that we write in symbolic form as

$$\mathcal{N}(< u >) = 0 . \tag{9.1}$$

In reality, however, since $\beta_D$ is not zero, this relation must be written as in (7.1), (7.7), a form that we indicate schematically as

$$\mathcal{N}(< u >) = \beta_D \mathcal{M}_0(< u >_1) . \tag{9.2}$$

The operator $\mathcal{M}_0$ depends also on $< u >$ and $\beta_D$ but is finite in the limit $\beta_D \to 0$. The presence of the conditional averages $< u >_1$ in $\mathcal{M}_0$ renders the formulation (9.2) not closed. The problem for the conditionally averaged fields $< u >_1$ is formulated in (6.8), (7.15). From the remarks given after Eq. (7.15), we deduce that its structure is of the form

$$\mathcal{N}(< u >_1) = \beta_D \mathcal{M}_1(< u >_2) . \tag{9.3}$$

The operator in the left-hand side is the same as in (9.2) but the solution $< u >_1$ must be sought in the domain exterior to the particle at $\mathbf{y}$. If, in place of this (exact) relation, we solve its approximation

$$\mathcal{N}(< u >_1) = 0 , \tag{9.4}$$

we incur in an error of order $\beta_D$ (or, at any rate, smaller than order one). Upon using the result of this operation in the right-hand side of (9.2), the resulting error is then of order higher than $\beta_D$ and can therefore be neglected for results accurate to order $\beta_D$ included. This procedure does not remove completely the mutual interaction of the particles. The particles interact through the average fields rather than directly.

The reason the single-particle problem (9.4) can be solved is that one only requires a local solution valid near the particle. The boundary conditions are the usual kinematic and dynamic ones at the particle surface while, for $|\mathbf{x} - \mathbf{z}| \to \infty$, we require (Hinch 1977)

$$\begin{aligned} < \mathbf{u}_C >_1 (\mathbf{z}, t | \mathbf{x}, \mathbf{w}) \to < \mathbf{u}_C > (\mathbf{x}, t) + (\mathbf{z} - \mathbf{x}) \cdot \nabla < \mathbf{u}_C > (\mathbf{x}, t) \\ + \frac{1}{2}(\mathbf{z} - \mathbf{x})(\mathbf{z} - \mathbf{x}) : \nabla\nabla < \mathbf{u}_C > (\mathbf{x}, t) , \end{aligned} \tag{9.5}$$

to maintain first-order accuracy in the ratio $a/L$. This condition may be justified by noting that the effect of the particle centered at $\mathbf{x}$ must disappear far from the particle.

On the basis of the framework just described, in our earlier papers we have derived rigorous dilute-limit closures for a number of problems. For example, for rigid spheres in potential flow (Zhang and Prosperetti 1994a), we find the following momentum equations that we write omitting the indication of averages for simplicity:

$$\begin{aligned}\rho_C\beta_C\left[\frac{\partial \mathbf{u}_C}{\partial t}+(\mathbf{u}_C\cdot\nabla)\mathbf{u}_C\right]+\beta_C\nabla p_C = \\ -\frac{1}{2}\,\rho_C\beta_D\left[\frac{\partial \mathbf{u}_C}{\partial t}+\mathbf{u}_C\cdot\nabla\mathbf{u}_C-\frac{\partial \mathbf{w}}{\partial t}-\mathbf{w}\cdot\nabla\mathbf{w}\right] \\ -\frac{1}{2}\,\rho_C\beta_D\,(\nabla\times\mathbf{u}_C)\times(\mathbf{w}-\mathbf{u}_C) \\ +\frac{1}{4}\,\rho_C\nabla\cdot\left\{\beta_D\left[(\mathbf{u}_C-\mathbf{w})^2\mathbf{I}-2(\mathbf{u}_C-\mathbf{w})(\mathbf{u}_C-\mathbf{w})\right]\right\} \\ -\frac{1}{4}\,\rho_C\nabla\left[\beta_D(\mathrm{Tr}\,\mathbf{M}_D)\right]+\beta_C\rho_C\mathbf{g}\,,\end{aligned} \tag{9.6}$$

$$\begin{aligned}\rho_D\beta_D\left[\frac{\partial \mathbf{w}}{\partial t}+(\mathbf{w}\cdot\nabla)\mathbf{w}\right]+\beta_D\nabla p_C = \\ \frac{1}{2}\,\rho_C\beta_D\left[\frac{\partial \mathbf{u}_C}{\partial t}+\mathbf{u}_C\cdot\nabla\mathbf{u}_C-\frac{\partial \mathbf{w}}{\partial t}-\mathbf{w}\cdot\nabla\mathbf{w}\right] \\ +\frac{1}{2}\,\rho_C\beta_D\,(\nabla\times\mathbf{u}_C)\times(\mathbf{w}-\mathbf{u}_C) \\ +\left(\rho_D+\frac{1}{2}\rho_C\right)\nabla\cdot(\beta_D\mathbf{M}_D)\,+\beta_D\rho_D\mathbf{g}\,.\end{aligned} \tag{9.7}$$

The first group of terms in the right-hand sides is the added mass interaction between the phases, and the following term the lift force.

As another example, when the Reynolds number for fluid-particle relative motion is very small, one finds (Zhang and Prosperetti 1997)

$$\begin{aligned}\beta_C\rho_C\left(\frac{\partial \mathbf{u}_C}{\partial t}+\mathbf{u}_C\cdot\nabla\mathbf{u}_C\right) = -\beta_C\nabla p_C+\beta_C\rho_C\mathbf{g} \\ +\beta_C\nabla\cdot(2\,\mu^*\mathbf{E}_m)+\frac{9\,\beta_D\,\mu_C}{2a^2}(\mathbf{w}-\mathbf{u}_C)-\frac{3}{4}\,\beta_D\mu_C\nabla^2\mathbf{u}_C \\ +\frac{3}{4}\mu_C\,\nabla^2[\beta_D(\mathbf{w}-\mathbf{u}_C)]+\nabla\cdot(\beta_C\mathbf{M}_C) \\ +3\mu_C\nabla\times\left[\beta_D\left(\mathbf{\Omega}-\frac{1}{2}\nabla\times\mathbf{u}_C\right)\right],\end{aligned} \tag{9.8}$$

$$\begin{aligned}\beta_D\rho_D\left(\frac{\partial\,\mathbf{w}}{\partial t}+\mathbf{w}\cdot\nabla\mathbf{w}\right) = -\beta_D\nabla p_C+2\,\mu_C\beta_D\nabla\cdot\mathbf{E}_C \\ -\frac{9\,\mu_C\beta_D}{2a^2}(\mathbf{w}-\mathbf{u}_C)+\frac{3}{4}\,\mu_C\beta_D\nabla^2\mathbf{u}_C+\nabla\cdot(\beta_D\mathbf{M}_D)+\beta_D\rho_D\mathbf{g}\,,\end{aligned} \tag{9.9}$$

where $\mu^*$ is the well-known Einstein effective viscosity of a dilute suspension of particles (see e.g. Batchelor 1967)

$$\frac{\mu^*}{\mu_C} = 1+\frac{5}{2}\,\beta_D+o(\beta_D)\,, \tag{9.10}$$

and $\mathbf{E}_m$, $\mathbf{E}_C$ are given by (7.10), (7.14) respectively. The equation (8.20) for the angular momentum of the particles, neglecting the rotational Reynolds stress (8.21) and collisional effects, is

$$(9.11) \qquad nJ\left[\frac{\partial \boldsymbol{\Omega}}{\partial t}+\mathbf{w}\cdot\nabla\boldsymbol{\Omega}\right] = 6\beta_D\mu_C\left(\frac{1}{2}\nabla\times\mathbf{u}_C-\boldsymbol{\Omega}\right)+n\mathbf{L}\,.$$

The disperse-phase Reynolds stress $\mathbf{M}_D$ cannot be determined internally by the theory without a specification of the initial conditions imposed on the particle probability distribution. This point was noted by Biesheuvel & Spoelstra (1989) who explicitly assumed that, at each position and time, the particle velocity probability distribution is strongly peaked around its local, instantaneous mean value. In this case $\mathbf{M}_D = 0$. A similar assumption – whether explicit or implicit – seems to be present in most of the previous work.

Results similar to those shown have been derived for spherical drops in Stokes flow (Zhang and Prosperetti 1997) and for spherical bubbles in potential flow (Zhang and Prosperetti 1994b). The method has also been applied to the convection-diffusion (energy) equation with similar results for particles at small Péclet number (Zhang and Prosperetti 1997).

**10. Numerical simulations.** We believe that the averaging procedure described in the previous sections provides an effective tool to achieve closure of the averaged equations on the basis of direct numerical simulations of some "canonical" multi-particle flows. We have given a first example of how this can be done for a simple potential flow case in an earlier paper (Zhang and Prosperetti 1994a). The method is as follows.

For a potential flow situation, $\mathcal{A}[\boldsymbol{\sigma}_C] = \mathcal{A}[-p_C\mathbf{I}]$, i.e.

$$(10.1)\ \mathcal{A}[-p_C\mathbf{I}] = -\frac{1}{\beta_D}\int d^3w P(\mathbf{x},\mathbf{w};t)\int_{|\mathbf{z}-\mathbf{x}|=a} dS_z\,(<p_C>_1 - <p_C>)\mathbf{n}.$$

The first crucial step is the choice of the averaged quantities in terms of which the theory will ultimately be phrased. In the paper cited we assume that all closure relations only depend on the phase average velocities, the disperse-phase volume fraction, and the continuous-phase average pressure. If we restrict consideration to the linear problem for a uniform mixture, then straightforward Continuum Mechanics arguments (Galilean invariance, time reversibility, isotropy, etc.) show that $\mathcal{A}$ must have the form

$$(10.2) \qquad \mathcal{A}[-p_C\mathbf{I}] = \frac{1}{2}\rho_C\beta_C\,C(\beta_D,\rho_D/\rho_C)\,\frac{\partial}{\partial t}(<\mathbf{u}_C> - \overline{\mathbf{w}})\,,$$

with $C$ a numerical coefficient whose dependence on the arguments $\beta_D$ and $\rho_D/\rho_C$ is unknown. This step reduces the determination of $\mathcal{A}$ to the calculation of the coefficient $C$, that can be shown to be related to the added mass coefficient introduced by several authors. On the basis of direct numerical simulations, we have calculated the integral (10.1) defining $\mathcal{A}$ and the accelerations in the right-hand side of (10.2). In this way it was then possible to calculate $C$ explicitly with the results shown in Fig. 1 of Zhang and Prosperetti (1994a).

This idea can be extended to more general situations. The starting point is a suitable guess concerning the functional dependence of the closure quantities (e.g. $\mathbf{M}_C$, $\mathcal{A}[\boldsymbol{\sigma}_C]$, $\mathcal{T}[\boldsymbol{\sigma}_C]$, etc.) on the "fundamental" average variables $\beta_D$, $< \mathbf{u}_C >$, $\overline{\mathbf{w}}$, $< p_C >$ (and possibly others, such as $\mathbf{M}_D$, for which new equations must be derived). In some cases, this functional dependence can be derived. In others, it may be suggested by fundamental arguments as before, supplemented by physical intuition, the results of approximate closures, and numerical experience with averaged equations. The next step is the actual numerical evaluation of the closure quantities according to their definitions, similar to (10.1), and of the mean variables. From these numerical results the postulated functional dependence can be made precise as in the previous example where the coefficient $C$ was determined in this way.

While the general idea is straightforward in principle, its actual implementation is far from trivial. For example, to calculate a continuous phase average $< f_C > (\mathbf{x}, t)$ according to the definition (5.1), one would have to run many simulations up to the time $t$, and average the values $f_C(\mathbf{x}, t; N)$ thus found discarding the simulations for which the point $\mathbf{x}$, at that time, happens to be in the disperse phase. By subdividing the region of interest into cells, a discretized approximation to $< f_C > (\mathbf{x}, t)$ can be constructed in this way. This approximation relies on the basic idea of the Monte Carlo evaluation of integrals according to which

$$\frac{1}{\mu(\mathcal{D})} \int_{\mathcal{D}} d\mathcal{D}\ \phi(\mathbf{X}) = \frac{1}{N_v} \sum_{j=1}^{N_v} \phi(\mathbf{X}_j), \tag{10.3}$$

where $\mathcal{D}$ is the integration domain, $\mu(\mathcal{D})$ its measure, and the integrand $\phi$ is evaluated at a set of $N_v$ points $\mathbf{X}_j \in \mathcal{D}$ chosen "at random" or, for faster convergence, according to a quasi-random rule (see e.g. Niederreiter 1992; Evans 1993).

The previous prescription runs into serious difficulties if the calculation of conditional averages such as $< f_C >_1 (\mathbf{x}, t | \mathbf{y}, \mathbf{w})$ is attempted, as all the values of $f_C(\mathbf{x}, t; N)$ generated in the simulations for which there isn't a particle at $(\mathbf{y}, \mathbf{w})$ must be discarded. Clearly, the amount of computation required for a reasonable discrete approximation to $< f_C >_1$ becomes astronomical and the evaluation of integrals such as those appearing in (5.5) or (5.7) impossible. It is primarily for this reason that we developed the "small particle" approximations (5.21), (5.24), (5.28), etc. The integrals appearing here are over a fixed particle, and can therefore be calculated for each particle in each simulation. Again by using a discretization of the domain of interest, a discrete approximation to $\mathcal{A}[\boldsymbol{\sigma}_C]$ or $\mathcal{A}'[\boldsymbol{\sigma}_C]$ etc. becomes feasible in this way.

The complexity of the calculation decreases substantially if consideration is limited to homogeneous suspensions as all average quantities become then spatially uniform. The standard way to deal with this situation numerically is to approximate the suspension by a finite domain with periodic boundary conditions which is equivalent to considering the whole space filled with copies of the fundamental domain. In view of the assumed homogeneity, two flows that differ by a rigid translation are both legitimate members of the ensemble, which implies that one can generate a reasonable statistics by running many (as opposed to a large number of) simulations. If $V$ is the volume of the fundamental domain, and if $< f_C >$ is independent of position, it can be calculated from its volume average and, using the definition (5.1) and interchanging the order of the integrations, we then have

$$\begin{aligned} < f_C > &= \frac{1}{V}\int_V d^3x < f_C > (\mathbf{x},t) \\ &= \frac{1}{N!\beta_C}\int d\mathcal{C}^N\, P(N;t)\,\frac{1}{V}\int_V d^3x\, f_C(\mathbf{x},t;N)\,\chi_C(\mathbf{x};N)\,. \end{aligned} \tag{10.4}$$

The integration over $\mathbf{x}$ is done either analytically as in Zhang and Prosperetti (1994a), or by a Monte Carlo method. The integration over the configurations is effected by averaging over many realizations, which, again, is a form of Monte Carlo integration. In principle this procedure should be followed for each permutation of the particle positions, with results then added and divided by $N!$. This step is evidently unnecessary as all permutations would give the same result. It is therefore consistent to simply ignore the factor $N!$ in (10.4).

A similar idea can be used for the particle averages starting from (4.1). In the uniform case, by averaging over the volume as before, we find

$$n\overline{g} = \frac{1}{N!}\int d\mathcal{C}^N\, P(N;t)\left[\frac{1}{V}\sum_{\alpha=1}^{N} g^\alpha(N,t)\right]. \tag{10.5}$$

The quantity in brackets is the mean over the particles, and the configuration integral is done by averaging over configurations; the factor $N!$ can again be ignored in practice.

The same prescription can be used for the calculation of $\mathcal{A}'[f_C]$ defined in (5.28). Indeed, upon substituting into (5.28) the conditional average $< f_C >_1$ according to the definition (5.3), we have

$$\beta_D\mathcal{A}'[f_C] = n\,\overline{\int_{|\mathbf{r}|=a} dS_r\, f_C(\mathbf{x}+\mathbf{r},t;N)\,\mathbf{n}}\ . \tag{10.6}$$

This expression has the form of a particle average and can therefore be calculated as in (10.5).

The calculation of the required averages for non-uniform suspensions is a much more complex matter. This is very unfortunate as, in order to develop a complete theory, it is imperative to deal with the non-uniform case. For example, again referring to the linear potential flow example used at the beginning of this section, it can be argued that the proper closure for $\mathcal{T}[-p_C\mathbf{I}]$ is:

$$\begin{aligned}(\mathcal{T}[-p_C\mathbf{I}])_{ij} = \rho_C a^2 \Big\{ & D_1 \left[\nabla\beta_D \cdot \frac{\partial}{\partial t}(<\mathbf{u}_C> - \overline{\mathbf{w}})\right]\delta_{ij} \\ & + \frac{1}{2} D_2 \left[\frac{\partial\beta_D}{\partial x_i}\frac{\partial}{\partial t}(<u_{Cj}> - \overline{w}_j) + \frac{\partial\beta_D}{\partial x_j}\frac{\partial}{\partial t}(<u_{Ci}> - \overline{w}_i)\right] \\ & + \left[D_3 \frac{\partial}{\partial t}(\nabla\cdot <\mathbf{u}_C>) + D_4 \frac{\partial}{\partial t}(\nabla\cdot\overline{\mathbf{w}})\right]\delta_{ij}\Big\}, \end{aligned} \tag{10.7}$$

where $D_k = D_k(\beta_D, \rho_D/\rho_C)$, $k = 1,2,3,4$. Clearly, none of these coefficients can be calculated on the basis of uniform particle distributions. Such derivative terms are likely to be of the greatest importance as it is by now recognized that there is little hope of developing realistic models of multiphase flows without at least second-order spatial derivatives (see e.g. Batchelor 1988).

We are in the process of developing an approach to this problem (Prosperetti and Marchioro 1996). The idea is that, if the postulated closure relations can indeed be considered as constitutive relations, then they should hold irrespective of the magnitude of the gradients. One can then generate a slightly non-uniform mixture and expand everything in the degree of non-uniformity. With this method, all the actual calculations are still carried out on a uniform suspension. The idea is similar to the well-known domain perturbation technique, where all actual calculations are carried out on the unperturbed domain.

**11. Turbulence.** It is not the purpose of this section to analyze the immensely complex problem of turbulence in multiphase flow, but only to point out how, in principle, turbulence is properly considered in the formalism described in this paper.

For the present purposes we think of turbulence as a situation in which a number (possibly infinite) of the fluid's degrees of freedom are not tracked deterministically. The quantities of interest will therefore be averaged over these degrees of freedom that we collectively denote by $\mathbf{Q} \equiv \{\mathbf{q}^a\}$, $a = 1, 2, \ldots$. The appropriate probability distribution will now have the form $\mathcal{P}(N, \mathbf{Q}; t)$, and the probability distribution for the particles alone that we have been using up to this point will be related to $\mathcal{P}$ by

$$P(N;t) = \int d\mathbf{Q}\, \mathcal{P}(N, \mathbf{Q}; t), \tag{11.1}$$

where the integration is over the entire range of $\mathbf{Q}$.

The definition of averaged quantities is formally as before, provided that the integration over $\mathbf{Q}$-space is added to that over the particle configurations. While one can now define several different conditional averages, the most useful ones differ from (5.3) only because of the extra $\mathbf{Q}$-integration. For example:

$$
\begin{aligned}
< f_{C,D} >_K (\mathbf{x}, t|K) \; &= \; \frac{1}{(N-K)!\beta_{C,D}^{(K)}} \\
&\times \int d\mathbf{Q} \int dC^{N-K} \; f_{C,D}(\mathbf{x}, t; N)\, \chi_{C,D}(\mathbf{x}; N)\, \mathcal{P}(N-K, \mathbf{Q}|K; t),
\end{aligned}
\tag{11.2}
$$

where

$$
\mathcal{P}(N-K, \mathbf{Q}|K; t) \; = \; \frac{\mathcal{P}(N, \mathbf{Q}; t)}{P(K; t)}, \tag{11.3}
$$

with

$$
P(K; t) \; = \; \frac{1}{(N-K)!} \int d\mathbf{Q} \int dC^{N-K} \mathcal{P}(N, \mathbf{Q}; t). \tag{11.4}
$$

Several of the relations given earlier in section 5.1 are formally unchanged with this extended averaging. In particular, the transport theorem (5.15) still holds, and so do the relations for the spatial gradients. As a consequence, the continuous phase continuity (7.1) and momentum (7.7) equations stand formally unchanged. One obvious conceptual point is, of course, that the Reynolds stress (7.6) now includes contributions from both the particles and the continuous-phase turbulence. However, the key point where $\mathbf{Q}$-averaging introduces a difference is in the conditionally-averaged continuity and momentum equations. For an incompressible fluid the former can be written down directly from (6.8) as

$$
\begin{aligned}
\frac{\partial \beta_C^{(1)}}{\partial t} + \nabla \cdot (\beta_C^{(1)} < \mathbf{u}_C >_1) \; &= \\
& \frac{\beta_C^{(1)}}{P(1; t)} \left\langle \Delta_1 \cdot \left[ \left( \ll \dot{\mathbf{w}}^1 \gg_{1,\mathbf{Q}} - \dot{\mathbf{w}}^1 \right) P(1; t) \right] \right\rangle_1 ,
\end{aligned}
\tag{11.5}
$$

where now, in place of (2.13),

$$
\ll \dot{\mathbf{w}} \gg_{1,\mathbf{Q}} = \frac{1}{(N-1)!} \int d\mathbf{Q} \int dC^{N-K} \; P(N-1, \mathbf{Q}|1; t)\, \dot{\mathbf{w}}. \tag{11.6}
$$

This quantity is evidently sensitive to the continuous-phase turbulence. The momentum equation will also differ in a similar way from the non-turbulent one (7.15). Since the conditionally averaged fields affect the unconditionally averaged equations for both phases through the terms $\mathcal{A}[\boldsymbol{\sigma}_C]$ etc., there is a mechanism here that accounts for the two-way interaction between particles and fluid.

It was argued before that, in the dilute laminar case, the difference $\ll \dot{\mathbf{w}} \gg_1 - \dot{\mathbf{w}}$ was $o(1)$ for $\beta_D \to 0$ because there were no fluctuations in the absence of particles. This statement does not evidently apply to $\ll \dot{\mathbf{w}} \gg_{1,\mathbf{Q}} - \dot{\mathbf{w}}$. We thus find an effect of turbulence even in the dilute limit.

**12. Conclusions.** This paper has focused on the technical aspects of an approach to the derivation of averaged equations for disperse flows. Some explicit results have been published (Zhang and Prosperetti 1994a, 1994b, 1997; Prosperetti and Zhang 1995, 1996; Prosperetti and Marchioro 1996), and several others are being worked out.

In our view, the method is interesting on several counts. In the first place, it enables one to deduce very readily and in a unified way exact results for a variety of problems involving dilute particle concentrations. Some examples have been given in section 9, and many more are contained in the references cited. Secondly – although we have not touched on this point – it can be used as a basis for the formulation of a variety of approximations, again in a unified way. Thirdly - and perhaps most importantly – it leads to a formulation of the closure problem in which the unknown quantities are expressed in a computable form. As we have shown in one case (Zhang and Prosperetti 1994a; see also Prospertti and Marchioro 1996), this statement is not only valid in principle, but can in practice lead to a solution of the closure problem. Work on several other cases is currently under way.

**Acknowledgments.** The author is grateful to D.Z. Zhang and M. Marchioro for several discussions on the subject matter of this paper. The work has been supported by NSF grant CTS-9521374 and DOE grant DE-FG02-89ER14043.

REFERENCES

[1] T.B. Anderson and R. Jackson, *A fluid mechanical description of fluidized beds*, *I & EC Fundamentals*, 6:527–539, 1967.

[2] G.K. Batchelor, *Sedimentation in a dilute dispersion of spheres*, *J. Fluid Mech.*, 52:245–268, 1972.

[3] G.K. Batchelor., *Transport properties of two-phase materials with random structure*, *Ann. Rev. Fluid Mech.*, 6:227–255, 1974.

[4] G.K. Batchelor, *A new theory of the instability of a uniform fluidized bed*, *J. Fluid Mech.*, 193:75–110, 1988.

[5] A. Biesheuvel and S. Spoelstra., *The added mass coefficient of a dispersion of spherical gas bubbles in liquid*, *Int. J. Multiphase Flow*, 15:911–924, 1989.

[6] A. Celmins, *Representation of two-phase flows by volume averaging*, *Int. J. Multiphase Flow*, 14:81–90, 1988.

[7] D.A. Drew, *Mathematical modeling of two-phase flow*, *Ann. Rev. Fluid Mech.*, 15:261–291, 1983.

[8] D.A. Drew and R.T. Jr. Lahey, *Analytical modeling of multiphase flow*, In Roco M.C., editor, *Particulate Two-Phase Flow*, pages 509–566, Butterworth-Heinemann, Boston, 1993.

[9] G. EVANS, *Practical Numerical Integration*, Wiley, Chichester, 1993.

[10] R.C. GIVLER, *An interpretation of the solid phase pressure in slow fluid-particle flows*, *Int. J. Multiphase flow*, 13:717–722, 1993.

[11] E.J. HINCH, *An averaged-equation approach to particle interactions in fluid suspension*, *J. Fluid Mech.*, 83:695–720, 1977.

[12] M. ISHII, *Thermo-Fluid Dynamic Theory of Two-Phase Flow*, Eyrolles, Paris, 1975.

[13] T.S. LUNDGREN, *Slow flow through stationary random beds and suspensions of spheres*, *J. Fluid Mech.*, 51:273–299, 1972.

[14] H. NIEDERREITER, *Random Number Generation and Monte Carlo Methods*, Society for Industrial and Applied Mathematics, Philadelphia, 1992.

[15] R.I. NIGMATULIN, *Spatial averaging in the mechanics of heterogeneous and dispersed systems*, *Int. J. Multiphase Flow*, 5:353–385, 1979.

[16] A. PROSPERETTI AND M. MARCHIORO, *Conduction in non-uniform composites*, In O. Manley and R. Goulard, editors, *Proceedings of the 14th Symposium on Energy Engineering Sciences*, Argonne National Laboratory, 1996, Report CONF-9605186.

[17] A. PROSPERETTI AND D.Z. ZHANG, *Finite-particle-size effects in disperse two-phase flows*, *Theor. Comput. Fluid Dynamics*, 7:429–440, 1995.

[18] A. PROSPERETTI AND D.Z. ZHANG, *Disperse phase stress in two-phase flow*, *Chem. Eng. Comm.*, 141-142:387–398, 1996.

[19] A.S. SANGANI, *A pairwise interaction theory for determining the linear acoustic properties of dilute bubbly liquids*, *J. Fluid Mech.*, 232:221–284, 1991.

[20] A.S. SANGANI AND A.K. DIDWANIA, *Dispersed-phase stress tensor in flows of bubbly liquids at large Reynolds numbers*, *J. Fluid Mech.*, 248:27–54, 1993.

[21] G.B. WALLIS, *The averaged Bernoulli equation and macroscopic equations of motion for the potential flow of a two-phase dispersion*, *Int. J. Multiphase Flow*, 17:683–695, 1991.

[22] D.Z. ZHANG, *Renormalization in the closure relations for particulate flows*, *in preparation*, 1997.

[23] D.Z. ZHANG AND A. PROSPERETTI, *Averaged equations for inviscid disperse two-phase flow*, *J. Fluid Mech.*, 267:185–219, 1994a.

[24] D.Z. ZHANG AND A. PROSPERETTI, *Ensemble phase-averaged equations for bubbly flows*, *Phys. Fluids*, 6:2956–2970, 1994b.

[25] D.Z. ZHANG AND A. PROSPERETTI, *Momentum and energy equations for disperse two-phase flows and their closure for dilute suspensions*, *Int. J. Multiphase Flow*, 23:425–453, 1997.

# BUBBLY FLOWS WITH GRAVITY AND VISCOSITY

PETER SMEREKA*

**Abstract.** A kinetic theory for bubbly flow is developed which accounts for weak viscous effects and gravity. It is shown that a uniform suspension of bubbles rising at a constant speed is a solution of the model. The calculated speed is in agreement with experimental findings. It is also shown that this steady solution is unstable.

**1. Introduction.** In this note the kinetic theory of an ideal bubbly flow developed by Russo & Smereka [1,2] will be extended to include gravity and weak viscous effects. There have been other investigations on this topic. Sangani & Didwania[3] and Smereka[4] performed numerical simulations of bubbly flows. Both studies found that in the presence of gravity and liquid viscosity the bubbles would tend to form pancake-shaped clusters perpendicular to gravity. van Wijngaarden and Kapteyn [5] examined concentration waves in bubbly flows accounting for gravity and liquid viscosity both experimentally and analytically. More recently, van Wijngaarden [6] calculated the steady rise speed of dilute bubble suspensions in good agreement with experiments.

**2. Equations of motion.** We will consider a collection of rigid, massless bubbles of identical sizes in an incompressible liquid. We shall consider bubbly flows with a moderate to high Reynolds number. This means the liquid velocity is well approximated by potential flow and the viscous effects occur in a thin layer near the bubble surface. The liquid velocity, $\boldsymbol{v}_\ell$, is therefore the gradient of a velocity potential, $\phi$, hence,

$$\boldsymbol{v}_\ell = \nabla\phi, \quad \boldsymbol{x} \in \mathcal{V},$$

where $\mathcal{V}$ is the domain occupied by the liquid. The velocity potential satisfies the following elliptic problem:

$$\nabla^2\phi = 0, \quad \boldsymbol{x} \in \mathcal{V} \quad \text{with} \quad \frac{\partial\phi}{\partial n} = \boldsymbol{u}_k \cdot \boldsymbol{n}, \quad \boldsymbol{x} \in S_k,\ k = 1, \ldots, N, \tag{2.1}$$

where $\boldsymbol{n}$ is the outward drawn normal from the liquid surface, $\boldsymbol{u}_k$ is the velocity of the $k$th bubble and $S_k$ is the surface of bubble $k$. The condition at infinity is $\nabla\phi = 0$ which corresponds to the zero volumetric flux frame of reference. Because (2.1) is linear in $\phi$, we can write

$$\phi = \sum_{k=1}^{N} \boldsymbol{u}_k \cdot \boldsymbol{\psi}_k, \tag{2.2}$$

* Department of Mathematics, University of Michigan, Ann Arbor, MI 48109.

where $\boldsymbol{\psi}_k$ is a vector valued function that depends only on the positions of the bubbles. The boundary conditions are then satisfied by taking

$$\frac{\partial \boldsymbol{\psi}_k}{\partial n} = \begin{cases} \boldsymbol{n} & \text{on the } k\text{th bubble} \\ 0 & \text{on the other bubbles.} \end{cases}$$

The total kinetic energy of the system is entirely contained in the liquid since the bubbles have no mass, therefore

$$T = \tfrac{1}{2}\rho_\ell \int_{\mathcal{V}} |\nabla\phi|^2 d\boldsymbol{x},$$

where $\rho_\ell$ is the density of the liquid. It can be shown [4] that

$$T = \tfrac{1}{2}\sum_{i,j} \boldsymbol{u}_i^T \mathsf{A}_{ij}\boldsymbol{u}_j, \quad \text{where} \quad \mathsf{A}_{ij} = \rho_\ell \int_{S_j} \boldsymbol{\psi}_i \boldsymbol{n}^T dS_j.$$

It follows from the definition of $\boldsymbol{\psi}_i$ that $\mathsf{A}_{ij}$ depends only on the positions of the bubbles; therefore $\mathsf{A}_{ij} = \mathsf{A}_{ij}(\boldsymbol{x}_1, \ldots, \boldsymbol{x}_N)$. The kinetic energy $T$ is the Lagrangian of a system of $3N$ degrees of freedom with coordinates $\{\boldsymbol{x}_1, \ldots, \boldsymbol{x}_N\}$ and velocities $\{\boldsymbol{u}_1, \ldots, \boldsymbol{u}_N\}$. The equations of motion are consequently given by the Euler-Lagrange equations:

$$\frac{d}{dt}\frac{\partial T}{\partial \boldsymbol{u}_k} - \frac{\partial T}{\partial \boldsymbol{x}_k} = \boldsymbol{Q}_k \quad \text{for} \quad k = 1 \text{ to } N \tag{2.3}$$

where $\boldsymbol{Q}_k$ is the external force on the $k$th bubble. This force arises from buoyant and viscous forces. To model the viscous drag force we shall use

$$12\pi\mu a\,[\boldsymbol{u}_k - \boldsymbol{v}(\boldsymbol{x}_k)]$$

where $\mu$ is the viscosity of the liquid, $a$ is the bubble radius, and $\boldsymbol{v}(\boldsymbol{x}_k)$ is the ambient liquid velocity at the center of the $k$th bubble. This formula is due to Levich. We shall consider bubbly fluids where the volumetric flux is zero; therefore the pressure drop between the inflow and outflow points must balance the weight of the liquid. Hence the average pressure gradient in the liquid is

$$-(1-\varepsilon_0)\rho_\ell g$$

where $g$ is the acceleration due to gravity and $\varepsilon_0$ is the average void fraction. This indicates that the buoyant force on a bubble is $\tau(1-\varepsilon_0)\rho_\ell g \boldsymbol{e}_x$ where $\tau$ is the volume of the bubble and $\boldsymbol{e}_x$ is unit vector in the $+x$-direction. Therefore, the external force on the $k$th bubble is

$$\boldsymbol{Q}_k = \tau(1-\varepsilon_0)\rho_\ell g \boldsymbol{e}_x - 12\pi\mu a\,[\boldsymbol{u}_k - \boldsymbol{v}(\boldsymbol{x})]\,. \tag{2.4}$$

It is useful to write the equations of motion in terms of the impulse of the bubble rather than its velocity. The impulse is given as

$$\boldsymbol{p}_k = \rho_\ell \int_{S_k} \phi \boldsymbol{n} dS = \sum_j \mathsf{A}_{jk}\boldsymbol{u}_k. \tag{2.5}$$

Substituting (2.5) into (2.3) we obtain

$$\begin{aligned}\dot{\boldsymbol{x}}_k &= \sum_i \mathsf{B}_{ki}\boldsymbol{p}_i \\ \dot{\boldsymbol{p}}_k &= -\tfrac{1}{2}\sum_{ij}\boldsymbol{p}_i^T \frac{\partial \mathsf{B}_{ij}}{\partial \boldsymbol{x}_k}\boldsymbol{p}_k + \boldsymbol{Q}_k\end{aligned}$$

with

$$\sum_k \mathsf{A}_{ik}\mathsf{B}_{kj} = I\delta_{ij},$$

where $I$ is the $3 \times 3$ identity matrix and $\delta_{ij}$ is the Kronecker delta.

There is no simple analytic expression for $\mathsf{B}_{ij}$ for a generic distribution of bubbles. Here we will make use of the point-bubble approximation. In this approximation the contribution of a single bubble to the velocity potential is given by a dipole field. This dipole field is an exact solution for a single bubble moving in an unbounded fluid. Our approximate expression for $\mathsf{B}_{ij}$ is

$$\mathsf{B}_{ij} = \begin{cases} \dfrac{2}{\rho_\ell \tau} I & i = j \\ -\dfrac{3}{\rho_\ell \tau}\left(\dfrac{a}{|\boldsymbol{r}_{ij}|}\right)^3 \left(I - 3\dfrac{\boldsymbol{r}_{ij}\boldsymbol{r}_{ij}^T}{|\boldsymbol{r}_{ij}|^2}\right) & i \neq j \end{cases} \tag{2.6}$$

where $\boldsymbol{r}_{ij} = \boldsymbol{x}_i - \boldsymbol{x}_j$. The equations of motion simplify when (2.6) is used. We can show (see [1]) that

$$\dot{\boldsymbol{p}}_k = -\frac{\partial}{\partial \boldsymbol{x}_k}(\boldsymbol{p}_k \cdot \boldsymbol{u}_k) + (1-\varepsilon_0)\rho_\ell g \boldsymbol{e}_x - 12\pi\mu a(\boldsymbol{u}_k - \boldsymbol{v}_k) \tag{2.7}$$

where

$$\boldsymbol{u}_k = \frac{2}{\tau\rho_\ell}\boldsymbol{p}_k + 3\boldsymbol{v}_k \quad \text{and} \quad \boldsymbol{v}_k = \tfrac{1}{3}\sum_{j\neq k}\mathsf{B}_{jk}\boldsymbol{p}_j.$$

**2.1. Dimensionless form.** We let $\boldsymbol{p}_k = \tau\rho_\ell U_\infty \boldsymbol{p}_k^*$, $\boldsymbol{x}_k = L\boldsymbol{x}_k^*$, and $t = (L/U_\infty)t^*$ where the asterisks denote dimensionless variables, $U_\infty$ is the steady rise speed of a single gas bubble in an unbounded fluid ($U_\infty = \rho_\ell a^2 g/9\mu$) and $L = (\tau/\varepsilon_0)^{1/3}$ ($L$ is proportional to the average inter-bubble distance). Substituting these variables into the equations of motion and dropping the asterisks one finds

$$\dot{\boldsymbol{p}}_k = -\frac{\partial}{\partial \boldsymbol{x}_k}(\boldsymbol{p}_k \cdot \boldsymbol{u}_k) + \gamma\left[(1-\varepsilon_0)\boldsymbol{e}_x - (\boldsymbol{u}_k - \varepsilon_0\boldsymbol{v}_k)\right] = \boldsymbol{F}_k \tag{2.8}$$

with

$$\boldsymbol{u}_k = 2\boldsymbol{p}_k + 3\varepsilon_0\boldsymbol{v}_k,$$

$$v_k = 3\sum_{j\neq k} \mathsf{B}(r_{jk})p_j, \qquad \mathsf{B}(r) = -\frac{1}{4\pi|r|^3}\left(I - 3\frac{rr^T}{|r|^2}\right),$$

and $\gamma = gL/U_\infty^2$ is the Froude number.

**3. Kinetic theory.** In this section we derive a self-consistent kinetic equation for the bubbles. We start with the conservation equation for the $N$-particle distribution function,

$$\frac{\partial}{\partial t}f^{(N)} + \sum_k \left[\frac{\partial}{\partial x_k}\cdot(u_k f^{(N)}) + \frac{\partial}{\partial p_k}\cdot(F_k f^{(N)})\right] = 0. \tag{3.1}$$

Following the procedure outlined in [1] and [2] we obtain a kinetic equation for $f(x, p, t)$, the number density of the bubbles in phase space. One finds

$$\frac{\partial f}{\partial t} + \frac{\partial}{\partial x}\cdot(uf) + \frac{\partial}{\partial p}\cdot(Ff) = Q(f, f) \tag{3.2}$$

where

$$u = 2p + 3\varepsilon_0 v$$

and

$$F = -3\varepsilon_0(\nabla_x v)^T p + \gamma\left[(1-\varepsilon_0)e_x + 2\varepsilon_0 v - 2p\right]$$

with

$$v = 3 ⨍ \mathsf{B}(x-y)j(y)dy = 3\mathcal{B}j - 2j.$$

Here $Q(f, f)$ is the collision operator, $\mathcal{B}$ is the divergence-free projection operator, and ⨍ denotes the principle value integral. The relation between $v$ and $\mathcal{B}$ can be found in [7], (also see [1]). We also note that $v$ is the ambient liquid velocity. The number density of bubbles in physical space is

$$n(x, t) = \int f(x, p, t)dp. \tag{3.3}$$

In view of our dimensionless form, a spatially homogeneous bubbly flow will have $n(x, \cdot) = 1$.

**4. Model predictions.** Consider (3.2) without collisions. It is clear since the divergence of $(u, F)$ is $-2\gamma$ that as $t \to \infty$, $f(x, p, t)$ will be concentrated on a set of zero measure. Further, since $f > 0$ and $\iint f dp dx$ is a constant then $f \to \infty$ on this set.

If we wish to find a steady spatially homogeneous solution then $f$ takes the form

$$f = \delta(p - p_\infty),$$

where $\delta$ is the Dirac delta function and

$$\boldsymbol{p}_\infty = \frac{(1-\varepsilon_0)}{2(1-2\varepsilon_0)}\boldsymbol{e}_x.$$

Notice that this implies that all the bubbles are moving with the same velocity, namely

$$\boldsymbol{u}_\infty = \frac{(1-3\varepsilon_0)(1-\varepsilon_0)}{(1-2\varepsilon_0)}\boldsymbol{e}_x = \left(1-2\varepsilon_0+\mathcal{O}(\varepsilon_0^2)\right)\boldsymbol{e}_x.$$

This is in agreement with experimental results, see Hetsroni[8].

A calculation reveals for the bubbly flows considered in [5] that $\gamma \gg \varepsilon_0$. Therefore after short time we would expect all of the bubbles to have approximately the same velocity, indicating the effect of collisions must be small; hence it is reasonable to neglect the collision term. A further simplification occurs if we consider one dimensional solutions. In one dimension, ignoring collisions, (3.2) simplifies to

$$\frac{\partial f}{\partial t} + \frac{\partial}{\partial x}(uf) + \frac{\partial}{\partial p}(Ff) = 0, \tag{4.1}$$

with

$$\begin{aligned} u &= 2p - 6\varepsilon_0 j, \\ F &= 6p\varepsilon_0\frac{\partial j}{\partial x} + \gamma(1-\varepsilon_0-2p+4\varepsilon_0 j), \\ j &= \int pf\,dp, \end{aligned}$$

where $f(x,p,t) = \int f(x,y,z,p,p_y,p_z,t)dp_y dp_z$ is assumed to be independent of $y$ and $z$.

Next, we consider (4.1) with spatially homogeneous initial conditions, that is $f(x,p,0) = f_0(p)$. It is easy to see that $f$ must remain spatially homogeneous and consequently (4.1) simplifies to

$$\frac{\partial f}{\partial t} + \frac{\partial}{\partial p}(Ff) = 0, \tag{4.2}$$

where $f = f(p,t)$ and $F = \gamma(1-\varepsilon_0-2p+4\varepsilon_0\int fpdp)$. The above equation can be solved using the method of characteristics and one finds the solution of the initial value problem to be

$$f(p,t) = e^{2\gamma t} f_0\left(j_0 + (p-p_\infty)e^{2\gamma t} + (p_\infty - j_0)e^{4\gamma\varepsilon_0 t}\right)$$

where $j_0 = \int f_0 p dp$. Therefore we see that

$$f(p,t) \to \delta(p-p_\infty) \quad \text{as} \quad t\to\infty,$$

provided $\varepsilon_0 < \frac{1}{2}$. This steady solution corresponds to a spatially homogeneous bubbly mixture with all the bubbles moving with the same speed, $u_\infty$. This solution is, however, unstable if one considers spatial variations. To see this we look for a weak solution of the form

$$f(x,p,t) = n(x,t)\delta(p - \overline{p}(x,t)).$$

The above will be a solution of (4.1) provided

$$\frac{\partial n}{\partial t} + \frac{\partial (n\overline{u})}{\partial x} = 0,$$

$$\frac{\partial \overline{p}}{\partial t} + \overline{u}\frac{\partial \overline{p}}{\partial x} = 6\varepsilon_0 \overline{p}\frac{\partial (n\overline{p})}{\partial x} + 2\gamma\left[p_\infty - \overline{p} + 2\varepsilon_0(n\overline{p} - p_\infty)\right],$$

where

$$\overline{u} = 2\overline{p}(1 - \varepsilon_0 n).$$

It is readily verified that this set of equations is ill-posed. This indicates that the steady solution $f = \delta(p - p_\infty)$, is unstable. This is consistent with the numerical simulations of [3] and [4]. These studies show that the bubbles will tend to form clusters as $t \to \infty$.

## REFERENCES


[1] G. Russo and P. Smereka, *"A Kinetic theory for bubbly flow I: Collisionless case"*, *SIAM J. Appl. Math.*, **56**, (1996).

[2] G. Russo and P. Smereka, *"A Kinetic theory for bubbly flow II: Fluid dynamic limit"*, *SIAM J. Appl. Math.*, **56**, (1996).

[3] A.S. Sangani and A.K. Didwania, *" Dynamic simulations of flows of bubbly liquids at large Reynolds numbers"*, *J. Fluid Mech.*, **250**, 307 (1993).

[4] P. Smereka, *"On the dynamics of bubbles in a periodic box"*, *J. Fluid Mech.*, **254**, 79 (1993).

[5] L. van Wijngaarden and C. Kapteyn, *"Concentration waves in dilute bubble/liquid mixtures"*, *J. Fluid Mech.*, **212**, 111 (1990).

[6] L. van Wijngaarden, *"The mean rise velocity of pairwise-interacting bubbles in liquid"*, *J. Fluid Mech.*, **251**, 55 (1993).

[7] P. Smereka, *"A Vlasov description of the Euler equation"*, Nonlinearity, **9**, 1361 (1996).

[8] G. Hetsroni (Editor), *"Handbook of Multiphase Systems"*, Hemisphere, Washington. D.C..

## IMA SUMMER PROGRAMS

1987 Robotics
1988 Signal Processing
1989 Robustness, Diagnostics, Computing and Graphics in Statistics
1990 Radar and Sonar (June 18 - June 29)
New Directions in Time Series Analysis (July 2 - July 27)
1991 Semiconductors
1992 Environmental Studies: Mathematical, Computational, and Statistical Analysis
1993 Modeling, Mesh Generation, and Adaptive Numerical Methods for Partial Differential Equations
1994 Molecular Biology
1995 Large Scale Optimizations with Applications to Inverse Problems, Optimal Control and Design, and Molecular and Structural Optimization
1996 Emerging Applications of Number Theory
1997 Statistics in Health Sciences
1998 Coding and Cryptography

## SPRINGER LECTURE NOTES FROM THE IMA:

*The Mathematics and Physics of Disordered Media*
Editors: Barry Hughes and Barry Ninham
(Lecture Notes in Math., Volume 1035, 1983)

*Orienting Polymers*
Editor: J.L. Ericksen
(Lecture Notes in Math., Volume 1063, 1984)

*New Perspectives in Thermodynamics*
Editor: James Serrin
(Springer-Verlag, 1986)

*Models of Economic Dynamics*
Editor: Hugo Sonnenschein
(Lecture Notes in Econ., Volume 264, 1986)

## The IMA Volumes in Mathematics and its Applications

*Current Volumes:*

1 **Homogenization and Effective Moduli of Materials and Media** J. Ericksen, D. Kinderlehrer, R. Kohn, and J.-L. Lions (eds.)
2 **Oscillation Theory, Computation, and Methods of Compensated Compactness** C. Dafermos, J. Ericksen, D. Kinderlehrer, and M. Slemrod (eds.)
3 **Metastability and Incompletely Posed Problems** S. Antman, J. Ericksen, D. Kinderlehrer, and I. Muller (eds.)
4 **Dynamical Problems in Continuum Physics** J. Bona, C. Dafermos, J. Ericksen, and D. Kinderlehrer (eds.)
5 **Theory and Applications of Liquid Crystals** J. Ericksen and D. Kinderlehrer (eds.)
6 **Amorphous Polymers and Non-Newtonian Fluids** C. Dafermos, J. Ericksen, and D. Kinderlehrer (eds.)
7 **Random Media** G. Papanicolaou (ed.)
8 **Percolation Theory and Ergodic Theory of Infinite Particle Systems** H. Kesten (ed.)
9 **Hydrodynamic Behavior and Interacting Particle Systems** G. Papanicolaou (ed.)
10 **Stochastic Differential Systems, Stochastic Control Theory, and Applications** W. Fleming and P.-L. Lions (eds.)
11 **Numerical Simulation in Oil Recovery** M.F. Wheeler (ed.)
12 **Computational Fluid Dynamics and Reacting Gas Flows** B. Engquist, M. Luskin, and A. Majda (eds.)
13 **Numerical Algorithms for Parallel Computer Architectures** M.H. Schultz (ed.)
14 **Mathematical Aspects of Scientific Software** J.R. Rice (ed.)
15 **Mathematical Frontiers in Computational Chemical Physics** D. Truhlar (ed.)
16 **Mathematics in Industrial Problems** A. Friedman
17 **Applications of Combinatorics and Graph Theory to the Biological and Social Sciences** F. Roberts (ed.)
18 ***q*-Series and Partitions** D. Stanton (ed.)
19 **Invariant Theory and Tableaux** D. Stanton (ed.)
20 **Coding Theory and Design Theory Part I: Coding Theory** D. Ray-Chaudhuri (ed.)
21 **Coding Theory and Design Theory Part II: Design Theory** D. Ray-Chaudhuri (ed.)
22 **Signal Processing Part I: Signal Processing Theory** L. Auslander, F.A. Grünbaum, J.W. Helton, T. Kailath, P. Khargonekar, and S. Mitter (eds.)

23 **Signal Processing Part II: Control Theory and Applications of Signal Processing** L. Auslander, F.A. Grünbaum, J.W. Helton, T. Kailath, P. Khargonekar, and S. Mitter (eds.)
24 **Mathematics in Industrial Problems, Part 2** A. Friedman
25 **Solitons in Physics, Mathematics, and Nonlinear Optics** P.J. Olver and D.H. Sattinger (eds.)
26 **Two Phase Flows and Waves** D.D. Joseph and D.G. Schaeffer (eds.)
27 **Nonlinear Evolution Equations that Change Type** B.L. Keyfitz and M. Shearer (eds.)
28 **Computer Aided Proofs in Analysis** K. Meyer and D. Schmidt (eds.)
29 **Multidimensional Hyperbolic Problems and Computations** A. Majda and J. Glimm (eds.)
30 **Microlocal Analysis and Nonlinear Waves** M. Beals, R. Melrose, and J. Rauch (eds.)
31 **Mathematics in Industrial Problems, Part 3** A. Friedman
32 **Radar and Sonar, Part I** R. Blahut, W. Miller, Jr., and C. Wilcox
33 **Directions in Robust Statistics and Diagnostics: Part I** W.A. Stahel and S. Weisberg (eds.)
34 **Directions in Robust Statistics and Diagnostics: Part II** W.A. Stahel and S. Weisberg (eds.)
35 **Dynamical Issues in Combustion Theory** P. Fife, A. Liñán, and F.A. Williams (eds.)
36 **Computing and Graphics in Statistics** A. Buja and P. Tukey (eds.)
37 **Patterns and Dynamics in Reactive Media** H. Swinney, G. Aris, and D. Aronson (eds.)
38 **Mathematics in Industrial Problems, Part 4** A. Friedman
39 **Radar and Sonar, Part II** F.A. Grünbaum, M. Bernfeld, and R.E. Blahut (eds.)
40 **Nonlinear Phenomena in Atmospheric and Oceanic Sciences** G.F. Carnevale and R.T. Pierrehumbert (eds.)
41 **Chaotic Processes in the Geological Sciences** D.A. Yuen (ed.)
42 **Partial Differential Equations with Minimal Smoothness and Applications** B. Dahlberg, E. Fabes, R. Fefferman, D. Jerison, C. Kenig, and J. Pipher (eds.)
43 **On the Evolution of Phase Boundaries** M.E. Gurtin and G.B. McFadden
44 **Twist Mappings and Their Applications** R. McGehee and K.R. Meyer (eds.)
45 **New Directions in Time Series Analysis, Part I** D. Brillinger, P. Caines, J. Geweke, E. Parzen, M. Rosenblatt, and M.S. Taqqu (eds.)

46 **New Directions in Time Series Analysis, Part II**
D. Brillinger, P. Caines, J. Geweke, E. Parzen, M. Rosenblatt, and M.S. Taqqu (eds.)
47 **Degenerate Diffusions**
W.-M. Ni, L.A. Peletier, and J.-L. Vazquez (eds.)
48 **Linear Algebra, Markov Chains, and Queueing Models**
C.D. Meyer and R.J. Plemmons (eds.)
49 **Mathematics in Industrial Problems, Part 5** A. Friedman
50 **Combinatorial and Graph-Theoretic Problems in Linear Algebra**
R.A. Brualdi, S. Friedland, and V. Klee (eds.)
51 **Statistical Thermodynamics and Differential Geometry of Microstructured Materials**
H.T. Davis and J.C.C. Nitsche (eds.)
52 **Shock Induced Transitions and Phase Structures in General Media** J.E. Dunn, R. Fosdick, and M. Slemrod (eds.)
53 **Variational and Free Boundary Problems**
A. Friedman and J. Spruck (eds.)
54 **Microstructure and Phase Transitions**
D. Kinderlehrer, R. James, M. Luskin, and J.L. Ericksen (eds.)
55 **Turbulence in Fluid Flows: A Dynamical Systems Approach**
G.R. Sell, C. Foias, and R. Temam (eds.)
56 **Graph Theory and Sparse Matrix Computation**
A. George, J.R. Gilbert, and J.W.H. Liu (eds.)
57 **Mathematics in Industrial Problems, Part 6** A. Friedman
58 **Semiconductors, Part I**
W.M. Coughran, Jr., J. Cole, P. Lloyd, and J. White (eds.)
59 **Semiconductors, Part II**
W.M. Coughran, Jr., J. Cole, P. Lloyd, and J. White (eds.)
60 **Recent Advances in Iterative Methods**
G. Golub, A. Greenbaum, and M. Luskin (eds.)
61 **Free Boundaries in Viscous Flows**
R.A. Brown and S.H. Davis (eds.)
62 **Linear Algebra for Control Theory**
P. Van Dooren and B. Wyman (eds.)
63 **Hamiltonian Dynamical Systems: History, Theory, and Applications**
H.S. Dumas, K.R. Meyer, and D.S. Schmidt (eds.)
64 **Systems and Control Theory for Power Systems**
J.H. Chow, P.V. Kokotovic, R.J. Thomas (eds.)
65 **Mathematical Finance**
M.H.A. Davis, D. Duffie, W.H. Fleming, and S.E. Shreve (eds.)
66 **Robust Control Theory** B.A. Francis and P.P. Khargonekar (eds.)
67 **Mathematics in Industrial Problems, Part 7** A. Friedman
68 **Flow Control** M.D. Gunzburger (ed.)

69 **Linear Algebra for Signal Processing**
A. Bojanczyk and G. Cybenko (eds.)
70 **Control and Optimal Design of Distributed Parameter Systems**
J.E. Lagnese, D.L. Russell, and L.W. White (eds.)
71 **Stochastic Networks** F.P. Kelly and R.J. Williams (eds.)
72 **Discrete Probability and Algorithms**
D. Aldous, P. Diaconis, J. Spencer, and J.M. Steele (eds.)
73 **Discrete Event Systems, Manufacturing Systems, and Communication Networks**
P.R. Kumar and P.P. Varaiya (eds.)
74 **Adaptive Control, Filtering, and Signal Processing**
K.J. Åström, G.C. Goodwin, and P.R. Kumar (eds.)
75 **Modeling, Mesh Generation, and Adaptive Numerical Methods for Partial Differential Equations** I. Babuska, J.E. Flaherty, W.D. Henshaw, J.E. Hopcroft, J.E. Oliger, and T. Tezduyar (eds.)
76 **Random Discrete Structures** D. Aldous and R. Pemantle (eds.)
77 **Nonlinear Stochastic PDEs: Hydrodynamic Limit and Burgers' Turbulence** T. Funaki and W.A. Woyczynski (eds.)
78 **Nonsmooth Analysis and Geometric Methods in Deterministic Optimal Control** B.S. Mordukhovich and H.J. Sussmann (eds.)
79 **Environmental Studies: Mathematical, Computational, and Statistical Analysis** M.F. Wheeler (ed.)
80 **Image Models (and their Speech Model Cousins)**
S.E. Levinson and L. Shepp (eds.)
81 **Genetic Mapping and DNA Sequencing**
T. Speed and M.S. Waterman (eds.)
82 **Mathematical Approaches to Biomolecular Structure and Dynamics**
J.P. Mesirov, K. Schulten, and D. Sumners (eds.)
83 **Mathematics in Industrial Problems, Part 8** A. Friedman
84 **Classical and Modern Branching Processes**
K.B. Athreya and P. Jagers (eds.)
85 **Stochastic Models in Geosystems**
S.A. Molchanov and W.A. Woyczynski (eds.)
86 **Computational Wave Propagation**
B. Engquist and G.A. Kriegsmann (eds.)
87 **Progress in Population Genetics and Human Evolution**
P. Donnelly and S. Tavaré (eds.)
88 **Mathematics in Industrial Problems, Part 9** A. Friedman
89 **Multiparticle Quantum Scattering With Applications to Nuclear, Atomic and Molecular Physics** D.G. Truhlar and B. Simon (eds.)
90 **Inverse Problems in Wave Propagation** G. Chavent, G. Papanicolau, P. Sacks, and W.W. Symes (eds.)
91 **Singularities and Oscillations** J. Rauch and M. Taylor (eds.)

92 **Large-Scale Optimization with Applications, Part I: Optimization in Inverse Problems and Design**
L.T. Biegler, T.F. Coleman, A.R. Conn, and F. Santosa (eds.)

93 **Large-Scale Optimization with Applications, Part II: Optimal Design and Control**
L.T. Biegler, T.F. Coleman, A.R. Conn, and F. Santosa (eds.)

94 **Large-Scale Optimization with Applications, Part III: Molecular Structure and Optimization**
L.T. Biegler, T.F. Coleman, A.R. Conn, and F. Santosa (eds.)

95 **Quasiclassical Methods**
J. Rauch and B. Simon (eds.)

96 **Wave Propagation in Complex Media**
G. Papanicolaou (ed.)

97 **Random Sets: Theory and Applications**
J. Goutsias, R.P.S. Mahler, and H.T. Nguyen (eds.)

98 **Particulate Flows: Processing and Rheology**
D.A. Drew, D.D. Joseph, and S.L. Passman (eds.)

## *FORTHCOMING VOLUMES*

1992–1992: *Control Theory*
Robotics

1995–1996: *Mathematical Methods in Materials Science*
Nonlinear Optical Materials
Numerical Methods for Polymeric Systems
Topology and Geometry in Polymer Science
1996 Summer Program: *Emerging Applications of Number Theory*

1996–1997: *Mathematics in High Performance Computing*
Algorithms for Parallel Processing
Evolutionary Algorithms
The Mathematics of Information Coding, Extraction and Distribution
Molecular Structure: Dynamics, Geometry and Topology
Structured Adaptive Mesh Refinement Grid Methods
Computational Radiology and Imaging: Therapy and Diagnostics
Mathematical and Computational Issues in Drug Design
Grid Generation and Adaptive Algorithms
Parallel Solution of Partial Differential Equations
1997 Summer Program: *Statistics in the Health Sciences*
Week 1: Genetics
Week 2: Imaging
Week 3: Diagnosis and Prediction
Weeks 4 and 5: Design and Analysis of Clinical Trials
Week 6: Statistics and Epidemiology: Environment and Health

1997–1998: *Emerging Applications for Dynamical Systems*
Numerical Methods for Bifurcation Problems